Marek Berezowski

Fractals, bifurcations and chaos in chemical reactors

AF138226

Marek Berezowski

Fractals, bifurcations and chaos in chemical reactors

Dynamics of chemical reactors

LAP LAMBERT Academic Publishing

Impressum / Imprint

Bibliografische Information der Deutschen Nationalbibliothek: Die Deutsche Nationalbibliothek verzeichnet diese Publikation in der Deutschen Nationalbibliografie; detaillierte bibliografische Daten sind im Internet über http://dnb.d-nb.de abrufbar.
Alle in diesem Buch genannten Marken und Produktnamen unterliegen warenzeichen-, marken- oder patentrechtlichem Schutz bzw. sind Warenzeichen oder eingetragene Warenzeichen der jeweiligen Inhaber. Die Wiedergabe von Marken, Produktnamen, Gebrauchsnamen, Handelsnamen, Warenbezeichnungen u.s.w. in diesem Werk berechtigt auch ohne besondere Kennzeichnung nicht zu der Annahme, dass solche Namen im Sinne der Warenzeichen- und Markenschutzgesetzgebung als frei zu betrachten wären und daher von jedermann benutzt werden dürften.

Bibliographic information published by the Deutsche Nationalbibliothek: The Deutsche Nationalbibliothek lists this publication in the Deutsche Nationalbibliografie; detailed bibliographic data are available in the Internet at http://dnb.d-nb.de.
Any brand names and product names mentioned in this book are subject to trademark, brand or patent protection and are trademarks or registered trademarks of their respective holders. The use of brand names, product names, common names, trade names, product descriptions etc. even without a particular marking in this works is in no way to be construed to mean that such names may be regarded as unrestricted in respect of trademark and brand protection legislation and could thus be used by anyone.

Coverbild / Cover image: www.ingimage.com

Verlag / Publisher:
LAP LAMBERT Academic Publishing
ist ein Imprint der / is a trademark of
OmniScriptum GmbH & Co. KG
Heinrich-Böcking-Str. 6-8, 66121 Saarbrücken, Deutschland / Germany
Email: info@lap-publishing.com

Herstellung: siehe letzte Seite /
Printed at: see last page
ISBN: 978-3-659-62127-7

Fractals, bifurcations and chaos in chemical reactors
Dynamics of chemical reactors

Marek Berezowski

For my Wife and my Universities

Table of contents

Introduction

The heart of most chemical plants is a chemical reactor. Basically there are two types of reactors: tubular (homogeneous or catalytic) and tank. Depending on the type, they are described by system of partial differential equations or by system of ordinary differential equations. Each of these models can generate complex solutions, including: multiple steady states, periodic oscillations, quasiperiodic oscillations, chaos. Analysis of this type of equations requires using sophisticated mathematical methods and complex numerical algorithms. In this study these phenomena and methods of analysis were presented. Particular attention is paid to the bifurcation problem and chaotic oscillations.

Different mathematical - numerical methods were presented which were used to solve above mention problems. The following concepts as: bifurcation, Lyapunov's exponent, Lyapunov's time and power spectrum were used for this purpose. In addition some case studies were presented in which modified parametric continuation method, particularly useful for determining the complex diagrams of steady states. The way of chaos crisis prediction was presented and optimization of reactor's process by relaxation method. Presentation of these phenomena on bifurcation diagrams and phase planes give, some times, fractal images. These results enables to evaluate sensitivity of reactor to changes environment conditions. Fractal dimension was used for this aim. This study is based on the author's own research cycle.

Chapter 1. Fractals from reactor models

As was shown, the solutions of models of recirculation tubular chemical reactors may be of a very complex dynamic character, including chaos [1-5]. These solutions are illustrated in a clear way by steady-state diagrams of various type, a.o. by the so-called Feigenbaum's diagram (Fig.1.1).

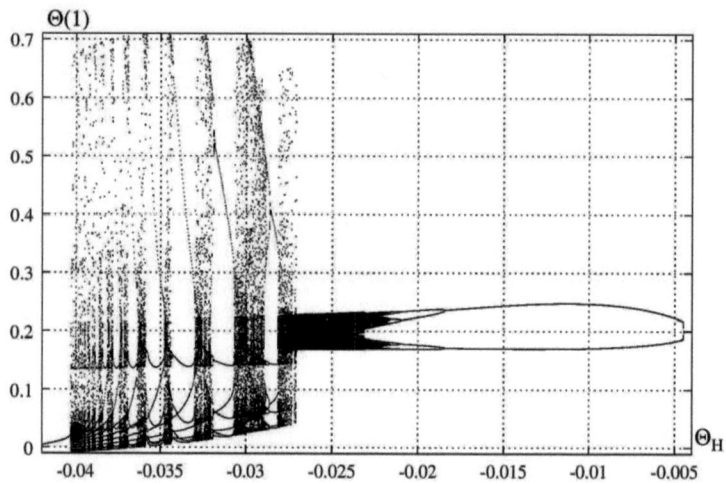

Fig. 1.1. *Complete Feigenbaum's diagram of the reactor with recycle of mass. The effect of the temperature dependence*

In the case when bifurcation parameter exceeds the value, which determines the entrance of the trajectory into the chaotic region, the system of diagram's branches has a fractal structure at this place [6]. In the present chaper the method of determination of the fractal dimension of Feigenbaum's tree, concerning a recirculation tubular chemical reactors, is demonstrated. Also the effect of initial value of state variables of rector, viz. degree of conversion α and temperature Θ, upon the value and convergence of model's solutions was investigated and fractal images were obtained. For the sake of computations a.o. the Mandelbrot's algorithm was applied [7,8]. In such a way both reactor's model with recirculation of mass and the model with thermal feedback were tested.

The mathematical model of a tubular non-adiabatic chemical reactor with recirculation of mass or heat comprises following balance equations:

mass:

$$\frac{\partial \alpha}{\partial \tau} + \frac{\partial \alpha}{\partial \xi} = \phi(\alpha, \Theta) \qquad (1.1)$$

heat:

$$\frac{\partial \Theta}{\partial \tau} + \frac{\partial \Theta}{\partial \xi} = \phi(\alpha, \Theta) + \delta(\Theta_H - \Theta) \qquad (1.2)$$

and algebraic balance equations, which constitute the boundary conditions resulting from the feedback:

$$\alpha(0, \tau) = xf\alpha(1, \tau) \qquad (1.3)$$

$$\Theta(0, \tau) = f\Theta(1, \tau) \qquad (1.4)$$

The reaction kinetics may be described by Arrhenius – type relationship:

$$\phi(\alpha, \Theta) = (1 - fx)Da(1 - \alpha)^n \exp(\gamma \frac{\beta \Theta}{1 + \beta \Theta}) \qquad (1.5)$$

In the case of recirculation of mass $x=1$ [1,3,4] and in the case of recirculation of heat $x=0$ [2,5].

The numerical analysis presented in papers [1-5] has demonstrated that there exist the regions of parameters values of both reactor models, i.e. with mass recirculation and with thermal feedback, with oscillation – type solutions, chaos including, these regions being of considerable magnitude. The character of these solutions is well exposed by Feigenbaum's tree in Fig.1.1. It concerns the reactor's temperature $\Theta(1)$. For the computations the following values of parameters were assumed: $Da=0.15$, $n=1.5$, $f=0.5$, $\gamma=15$, $\beta=2$, $\delta=3$. It is likely to imagine that the structure of this tree, based on the scenario of doubling the period, resembles the constructions of Cantor's set [6]. There is, however, a difference, because in Cantor's set the ratio of segment's division is strictly defined, being equal to $r=1/3$, whereas in the case of diagram from Fig.1.1 this ratio is different for different bifurcation points. The fractal dimension D of Cantor's set, determined from Kolmogorov's definition

$$2r^D = 2(1/3)^D = 1 \qquad\qquad (1.6)$$

equals $D=0.630929753....$. In the case of Feigenbaum's diagram from Fig.1.1 the individual branches disperse according to two different division rations r_1 and r_2, and not according to one only. It is easy to demonstrate (Appendix A1 of the chaper) that Kolmogorov's dimension D may be, in this case, determined from the relationship

$$r_1^D + r_2^D = 1 \qquad\qquad (1.7)$$

This quantity should be determined at the Feigenbaum's point, i.e. for the value of parameter Θ_H, which – when exceeded – leads to the entry of trajectory into the chaotic region. In the analysed example, approximately, $\Theta_H = -0.021739$. In Fig.1.2 the enlargement of a very small fragment of Fig.1.1 is shown.

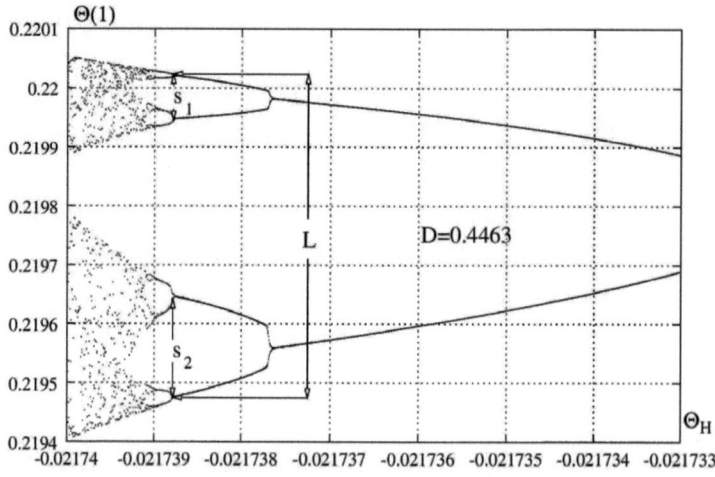

Fig.1.2. *Fragment of diagram from Fig.1.1*

Both division ratios were determined as

$$r_1 = \frac{s_1}{L}; \quad r_2 = \frac{s_2}{L} \tag{1.8}$$

which, after applying Eq.(1.7), yields the fractal division $D=0.4463$. In order to check the correctness of the obtained result, the enlargement of a fragment of Feigenbaum's diagram, this time referring to the reactor's conversion degree $\alpha(1)$, is given in Fig.1.3.

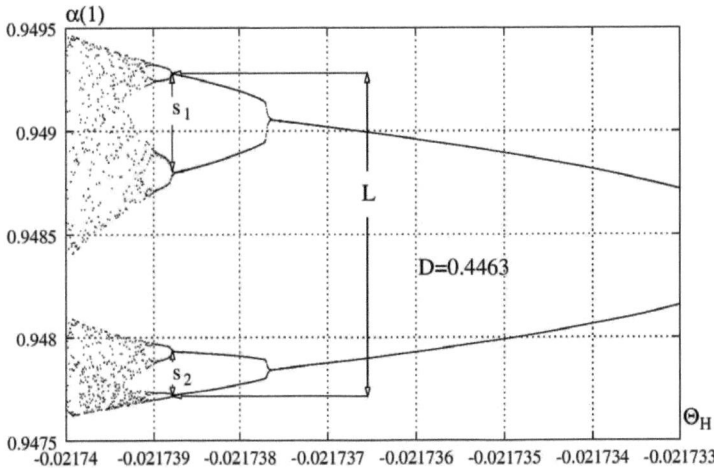

Fig.1.3. *Fragment of Feigenbaum's diagram of the reactor with recycle of mass. The effect of degree of conversion*

It is clearly seen that the obtained leaves, after change of scale, are identical with the leaves in Fig.1.2 (they overlap). The upper leaf from Fig.1.2 is proportional to the lower one from Fig.1.3, and the upper leaf from Fig.1.3 is proportional to the lower one from Fig.1.2. The change of place as above has no qualitative meaning and in each case the same value of dimension D is obtained from Eq.(1.7). In this chaper two types of fractal images, obtained from the mathematical models of recirculation tubular chemical reactors are presented. In order to determine them, two different algorithms were used.

Algorithm 1 of the generation of fractal images

Generating a discrete time series of state variables of the reactor $\alpha(1,k)$ and $\Theta(1,k)$ [1,3-5] a finite number of recurrent computational steps N was assumed, after which points with the coordinates $\{\alpha(1,0), \Theta(1,0)\}$ were plotted on the diagram. The shade of plotted point depended on the interval, in which found itself the value of $\Theta(1,N)$, i.e. the value of temperature in the last, N-th, iterative step (in oryninale is the colour [7,8]). In such a way a map of initial conditions, indicating their effect upon the value of solution, has been constructed. Assuming the values of parameters of reactor with recirculation of mass, as previously, and $\Theta_H = -0.023$, one has obtained image as in Fig.1.4. The number of steps of the recurrence process N=200. It is likely to imagine that Fig.1.4 resembles Saturn's rings.

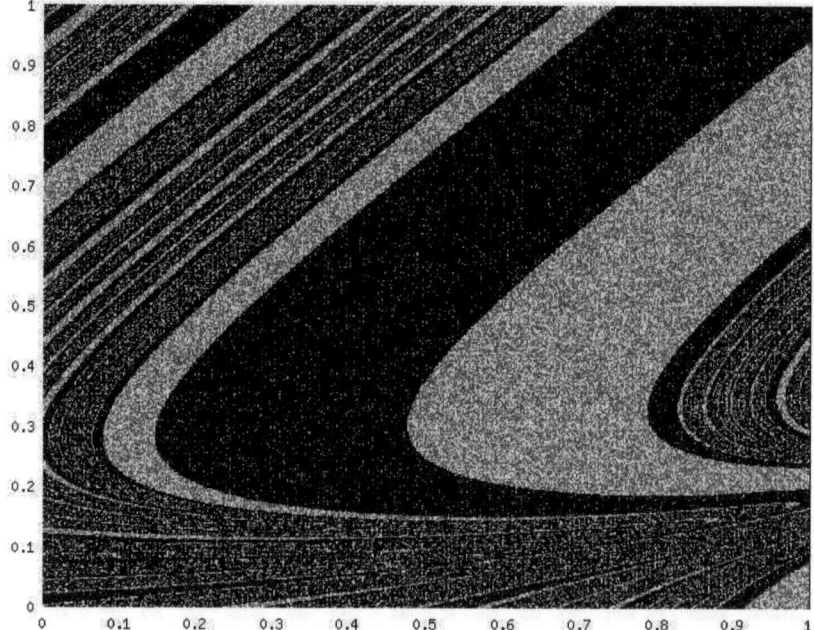

Fig.1.4. *Effect of initial conditions on the value of the solution of the model of reactor with the recycle of mass*

It is worth mentoring that individual bands in Fig.1.4 display a clearly outlined granular structure. This part bears witness to the sensibility of solutions with respect to initial conditions. It is interesting that within a given band one of the shade is dominating (in original is the colour [7,8]).

Algorithm 2 of the generation of fractal images of reactor

In this case, for the generation of fractal image the method of Mandelbrot has been used. Viz., it has been assumed that the state variables of reactor: conversion degree α and dimensionless temperature Θ, are complex numbers. From the physical point of view this assumption is, obviously, unjustified. This results from the fact that physical variables are of physical character exclusively. Nonetheless it has been proved that the models of recirculation tubular chemical reactors generate very interesting fractal structures for complex variables

$$\alpha = \alpha_r + i\alpha_i; \quad \Theta = \Theta_r + i\Theta_i \qquad (1.9)$$

what is interesting from mathematical viewpoint.

The following algorithm has been assumed. For fixed values $\alpha(0,0)$ the map of initial conditions in the coordinate system $\{\Theta_r(1,0), \Theta_i(1,0)\}$ was constructed. The shade of a point on the map depended on the number of iterative steps k, after which the length of the vector $|\Theta(1,k)|$ exceeded the given value of ε (in original is the colour [7,8]). This means that the computations were terminated, when

$$|\Theta(1,k)| \geq \varepsilon \qquad (1.10)$$

If the absolute value $|\Theta(1,k)|$ exceeded ε already in the first iterative step, no point was plotted on the diagram. In other words, the shades on the map correspond with the convergence rate of the numerical process. In such a way, changing successively $\Theta_r(1,0)$ and $\Theta_i(1,0)$ the fractal structures, as those on presented figures, were obtained. So, Fig.1.5 shows a basic fractal image of solutions of the model with thermal feedback ($x=0$).

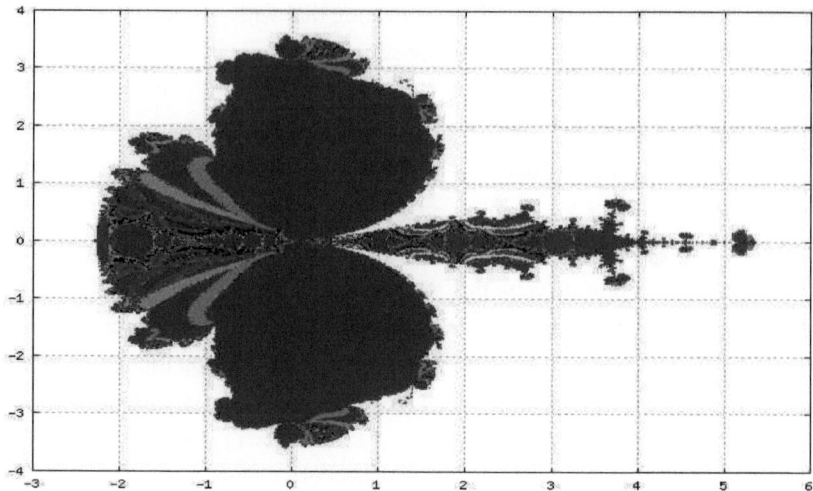

Fig.1.5. *Effect of initial conditions on the convergence of solutions of the model of reactor with the recycle of heat. General set*

In this case the value of $\alpha(0,0)$ results from the nature of the process and is equal to $\alpha_r(0,0) = 0$, $\alpha_i(0,0) = 0$ [5]. For the sake of computations the following values of parameters: Da=0.15, n=1, f=0.3, γ=10, β=1.4, $\Theta_H = -0.048$, N=200, ε=5 were assumed. A fragment of the set from Fig.1.5 is presented in Fig.1.6.

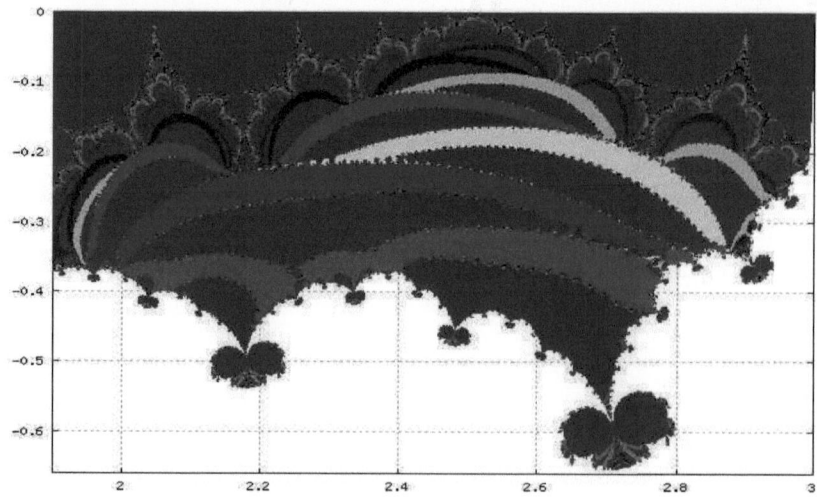

Fig.1.6. *Fragment of Fig.1.5*

13

There are clearly visible regions resembling the form of a basic set, which bears witness to the fractal structure of the obtained solutions. The successive figures, 1.7 and 1.8, present fragments of the set from Fig.1.6.

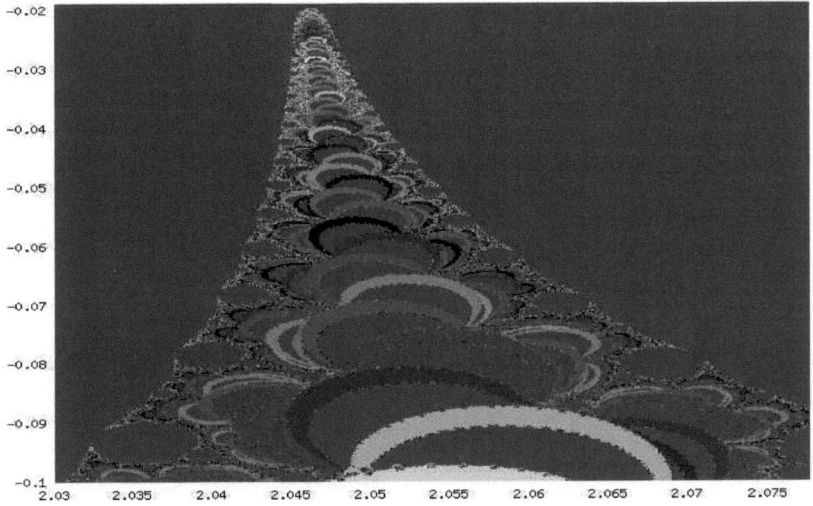

Fig.1.7. *Fragment of Fig.1.6*

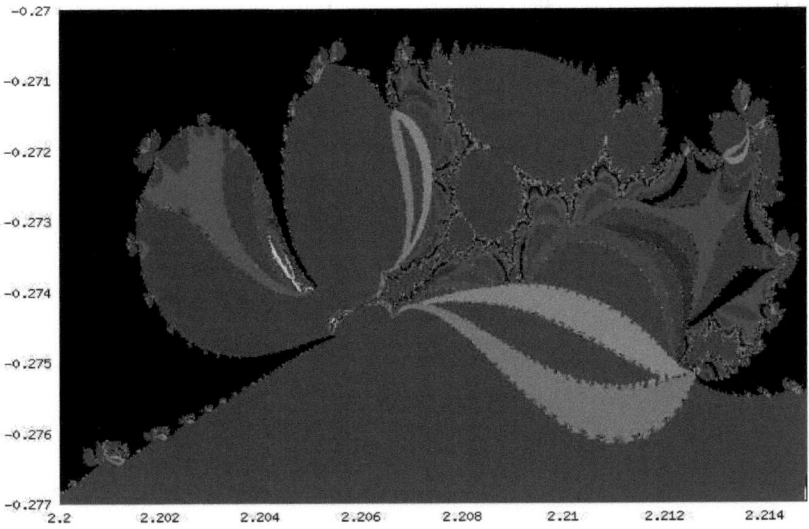

Fig.1.8. *Fragment of Fig.1.6*

Also here the elements similar to the basic set are distinct. In Fig.1.9 the basic fractal image of the solutions of the model of reactor with recirculation of mass (x=1) is shown [1,3,4].

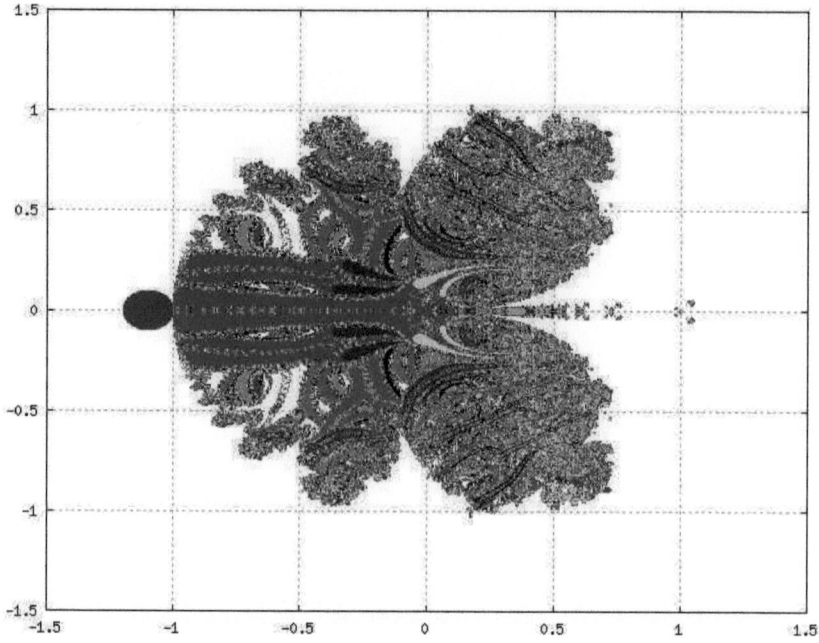

Fig.1.9. *Effect of initial conditions on the convergence of solutions of the model of reactor with the recycle of mass. General set*

In this case the values $\alpha_r(0,0)$=0.45, $\alpha_i(0,0)$=0, $\Theta_H = -0.033$, N=200 and ε=5 were assumed. The values of remaining parameters – as previously. The portions of the set from Fig.1.9 are presented in Figs 1.10 and 1.11.

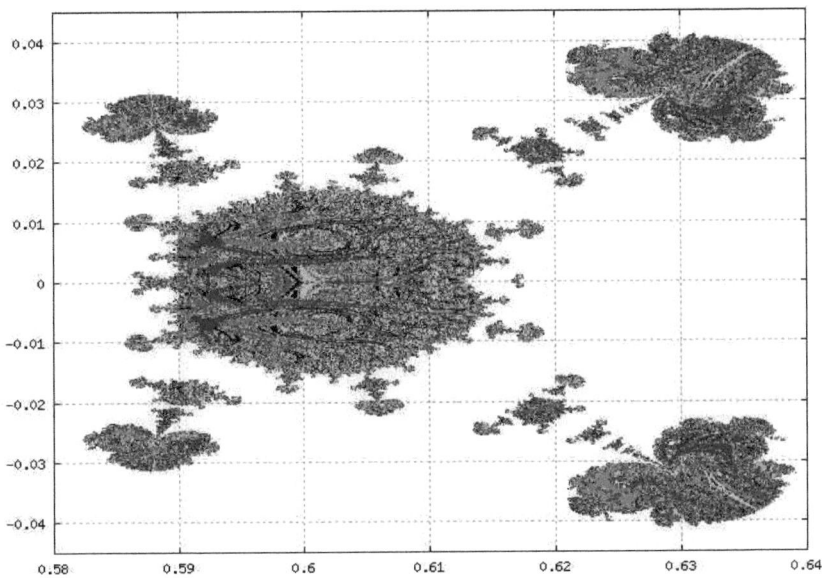

Fig.1.10. *Fragment of Fig.1.9*

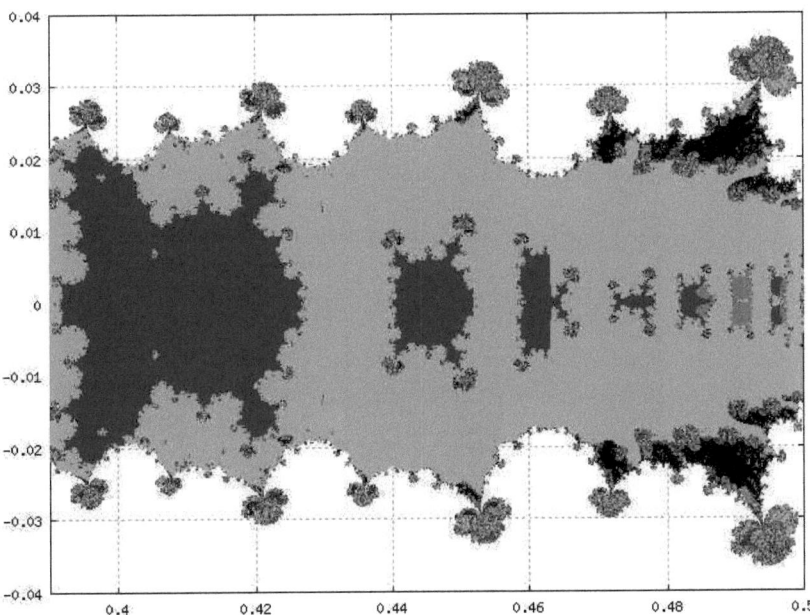

Fig.1.11. *Fragment of Fig.1.9*

Similarly as in the case of recirculation of heat also here the forms of the basic set are clearly visible. Fig.1.12 and 1.13 indicate the enlargement of the set from Fig.1.10, whereas Fig.1.14 – the enlargement of the fragment of the set from Fig.1.11.

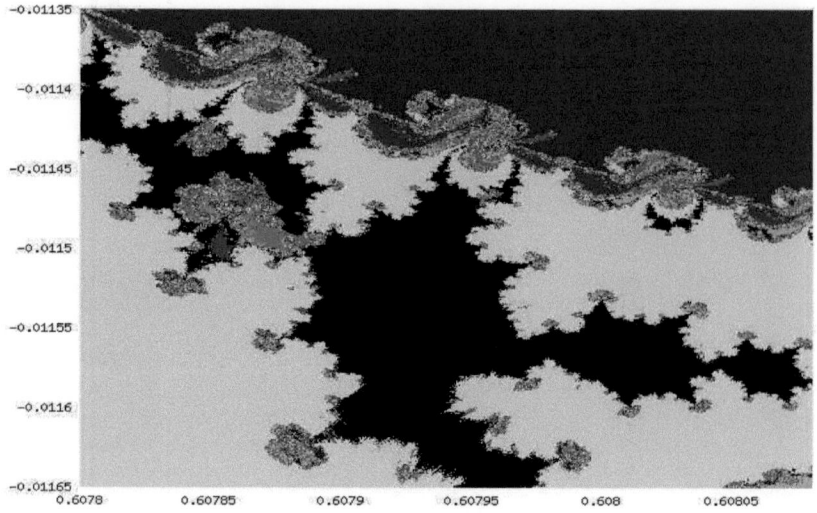

Fig.1.12. *Fragment of Fig.1.10*

Fig.1.13. *Fragment of Fig.1.10*

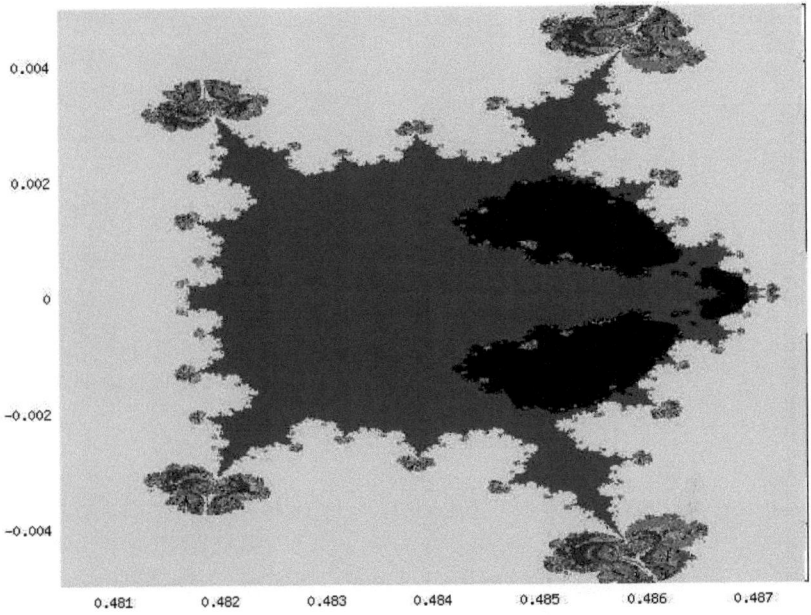

Fig.1.14. *Fragment of Fig.1.11*

Comparing the images obtained from the model of reactor with thermal feedback with those obtained from model of reactor with recirculation of mass one observes that the structure of the former ones is more uniform. Viz., the individual bands are of the same shade (compare Fig.1.5). In the case of the reactor with recycle the shades are more strongly mixed (compare Fig.1.9), which bears witness to a great turbulence of the process (in original are colours [7,8].

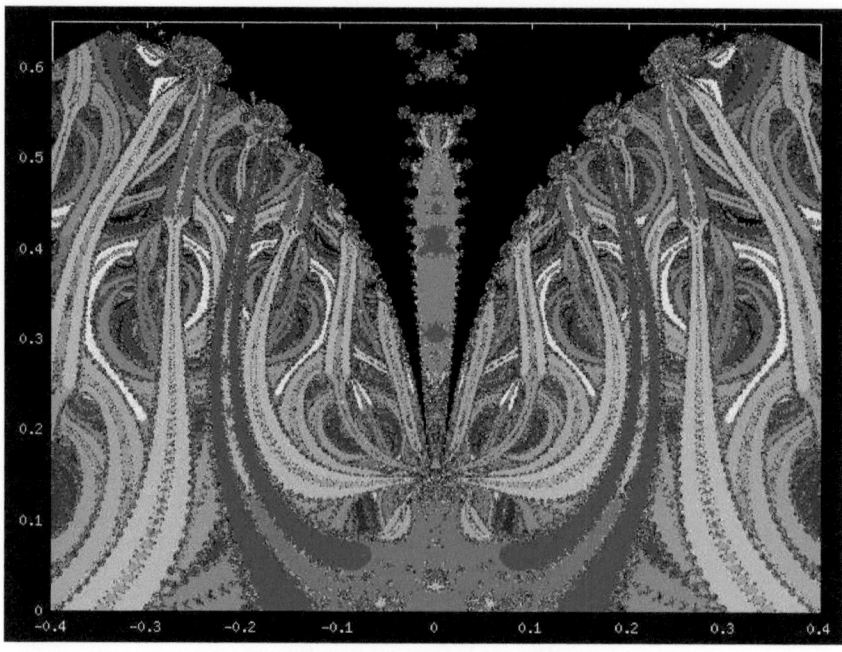

Fig. 1.15. *Fragment of Fig. 1.9*

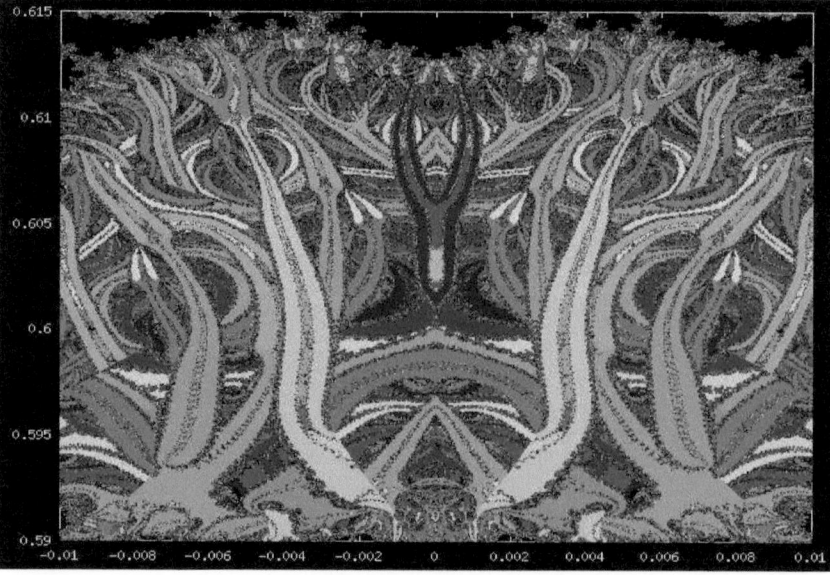

Fig. 1.16. *Fragment of Fig. 1.15*

Appendix A1. Fractal dimension of a set with two different division ratios

Let us divide the closed segment $[0,1]$ into three different parts, according to the division ratios r_1 and r_2 and remove the central part, leaving the boundary points. Let's do the same with two segments obtained $[0,r_1]$ and $[1-r_2,1]$, the same with four segments etc. (Fig.A1).

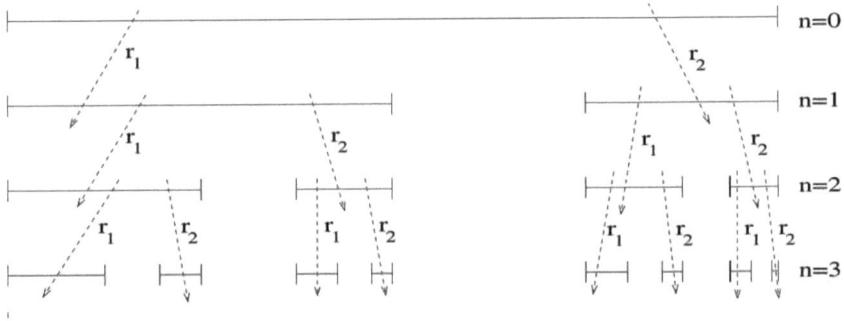

Fig.A1. *Construction of Cantor's set with two division ratios*

In the limit a set analogous to Cantor's set is obtained. Basing on Kolmogorov's definition of the fractal dimension D, it is easy to notice that in the case of two different division ratios, r_1 and r_2, this dimension may be determined, in *n-th* step of above algorithm, from Newton's binomial

$$1 = \sum_{k=0}^{n} \binom{n}{k} \left(r_1^{n-k} r_2^{k} \right)^{D} = \left(r_1^{D} + r_2^{D} \right)^{n} \Rightarrow r_1^{D} + r_2^{D} = 1 \qquad (A1.1)$$

It follows from (A.1) that D, similarly as in case of Cantor's set, does not depend on n, hence it is the general dimension of the set. The length of this set equals in the limit:

$$\lim_{n \to \infty} (r_1 + r_2)^{n} = 0 \qquad (A1.2)$$

References

1. E.W. Jacobsen, M. Berezowski. *Chaotic dynamics in homogeneous tubular reactors with recycle.* Chem.Engng Sci. **53**, 4023-4029 (1998).
2. M. Berezowski, P. Ptaszek, E.W. Jacobsen. *Dynamics of heat-integrated pseudohomogeneous tubular reactors with axial dispersion.* Chem.Engng Process, **39**, 181-188 (2000).
3. M. Berezowski. *Spatio-temporal chaos in tubular chemical reactors with the recycle of mass.* Chaos,Solitons&Fractals **11**, 1197-1204 (2000).
4. M. Berezowski. *Effect of delay time on the generation of chaos in continuous systems. One-dimensional model. Two-dimensional model - tubular chemical reactor with recycle.* Chaos,Solitons&Fractals **12**, 83-89 (2001).
5. M. Berezowski, A. Grabski. *Chaotic and non-chaotic mixed oscillations in a logistic systems with delay and heat-integrated tubular chemical reactor.* Chaos,Solitons&Fractals **14**/1, 97-103 (2002).
6. Peitgen HO, Jürgen H, Saupe D. *Fractals for the classroom.* New York: Springer (1992).
7. M. Berezowski, *Fractal solutions of recirculation tubular chemical reactors,* Chaos, Solitons&Fractals, **16**, 1-12, 2003.
8. M. Berezowski, *Fractals galery,* http://c504c.skroc.pl

Chapter 2. Fractal character of basin boundaries

Tubular chemical reactors with mass recycle are commonly applied in industrial processes. Due to feedback caused by recycle, complex static and dynamic phenomena may occur in the reactors (Luss & Amundson, 1967; Reilly & Schmitz 1966; Reilly & Schmitz 1967) including chaos (Jacobsen & Berezowski, 1998; Berezowski, 2000; Berezowski 2001). One of such phenomena is the steady states multiplicity of the reactor with the same values of its parameters. This concerns both static (stationary) and dynamic states, and, in particular, different types of oscillations. The set of initial conditions that tend asymptotically towards a given attractor as times goes to plus infinity is known as the basin of attraction of the attractor. The boundary separating the basins of attraction of different attractors is known as the basin boundary. The basin boundary may consist of specific trajectories in the phase space. However, the basin boundary may have a fractal structure (Nusse & Yorke, 1994). The discussed phenomenon may also be observed in Mandelbrot's set, where the basins separating particular attractors have a fractal nature. If so, the separation of the steady states (attractors) occurs chaotically, which, from the technological point of view, is a disadvantage. Even a small change in the values of temperature or flux concentration may lead to drastic change of the state of the reactor as a whole, for example, to the transfer from the stabilized stationary state into the periodic oscillation state, or from the periodic oscillation state to the chaotic one, etc. Consequently, from the practical point of view, the system becomes unpredictable. The scope of the chapter is to discuss the possibility of the occurrence of this phenomenon in chemical tubular reactors with recycle (Berezowski, 2006). Exemplary separation areas were designated, and their fractal dimensions and entropy determined. The analysis was conducted on a homogeneous non-adiabatic tubular chemical reactor with mass recycle. The mathematical model of the corresponding balances of mass and heat is constituted by the following equations:

$$\frac{\partial \alpha}{\partial \tau} + \frac{\partial \alpha}{\partial \xi} = (1 - f)\phi(\alpha, \Theta) \tag{2.1}$$

$$\frac{\partial \Theta}{\partial \tau} + \frac{\partial \Theta}{\partial \xi} = (1 - f)\phi(\alpha, \Theta) + (1 - f)\delta(\Theta_H - \Theta) \tag{2.2}$$

where the degree of conversion α and dimensionless temperature Θ are defined as:

$$\alpha = \frac{C_{A0} - C_A}{C_{A0}}; \quad \Theta = \frac{T - T_0}{\beta T_0}. \tag{2.3}$$

Function ϕ describing the kinetics of the chemical reaction has the following form:

$$\phi = Da(1 - \alpha)^n \exp\left(\gamma \frac{\beta \Theta}{1 + \beta \Theta}\right) \tag{2.4}$$

whereas the boundary conditions resulting from the recycle assume, accordingly, the following forms:

$$\alpha(0, \tau) = f\alpha(1, \tau); \quad \Theta(0, \tau) = f\Theta(1, \tau). \tag{2.5}$$

Different static and dynamic solutions may be derived from the simulation of the above model. They may involve multiple stationary states, limit cycles, quasiperiodic orbits or chaotic ones. Such results were discussed in: (Jacobsen &

Berezowski, 1998; Berezowski, 2000; Berezowski, 2001). The case presented on the bifurcation diagram in Fig.3 studied in (Berezowski, 2000) is particularly interesting. The chaotic areas are separated by multiple windows of periodic solutions. One such window occurs for $\Theta_H = -0.02875$, see Fig.2.1 below.

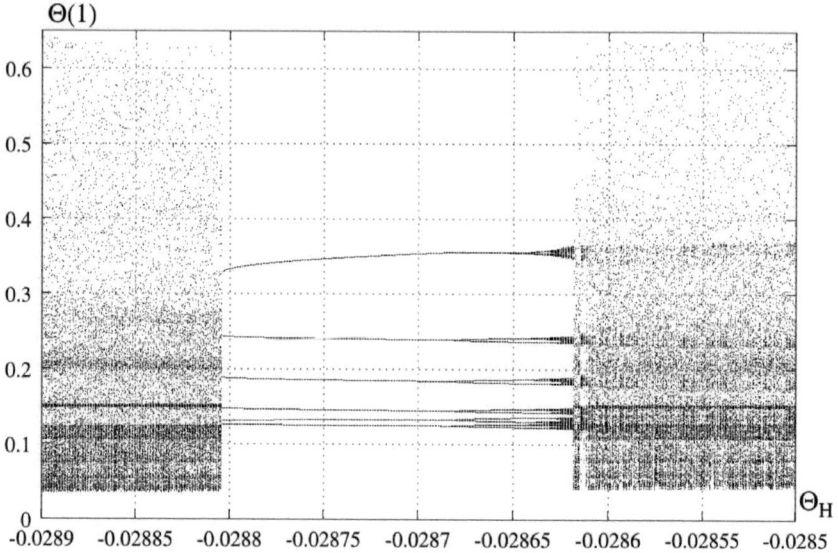

Fig.2.1. *Fragment of the bifurcation diagram of the reactor - window with a periodic solution*

Depending on the initial values of the temperature or flux concentration, this window may apply to a periodic solution (in this case, the hexa-periodic solution – Fig.2.1) or to a chaotic one (Fig.2.2).

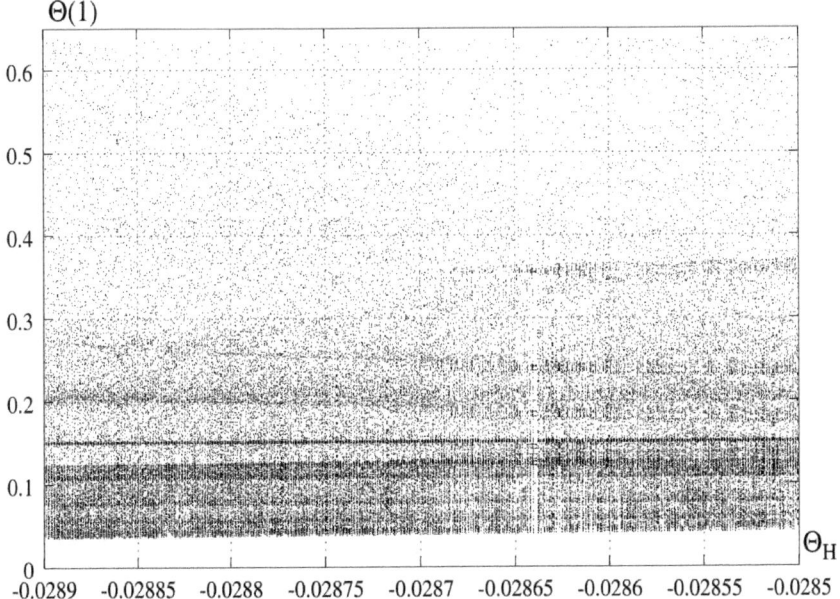

Fig. 2.2. *Fragment of the bifurcation diagram of the reactor - window with a chaotic solution*

Therefore, when $\Theta_H = -0.02875$ the system (2.1)-(2.5) has two attractors. The choice of initial condition determines which basin of attraction a given trajectory starts in and consequently the attractor to which it will evolve (Jaconsen & Berezowski, 1998; Berezowski, 2000). One of such attractors has the form of a hexa-periodic orbit, whereas the other one is a strange attractor. In the course of the numerical simulation of the above mentioned model (2.1-2.5) a map of the dynamics of the reactor was obtained- see Fig.2.3.

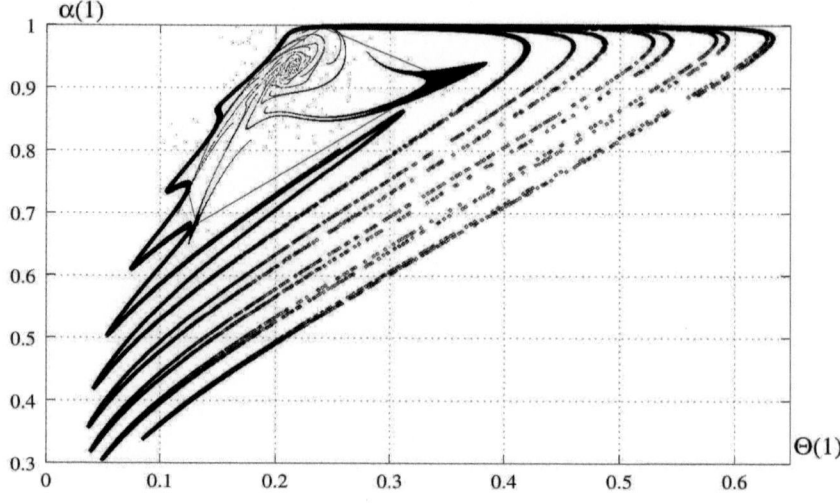

Fig.2.3. *Map of the dynamics of the reactor*

About 100000 grid points were used to designate the map $[\alpha(1), \Theta(1)]$. The big dots represent the strange attractor. The white area is the area of its attraction. The continuous line marks the stable hexa-periodic orbit and the set of small dots is the area of its attraction. This area is shown individually in Fig.2.4.

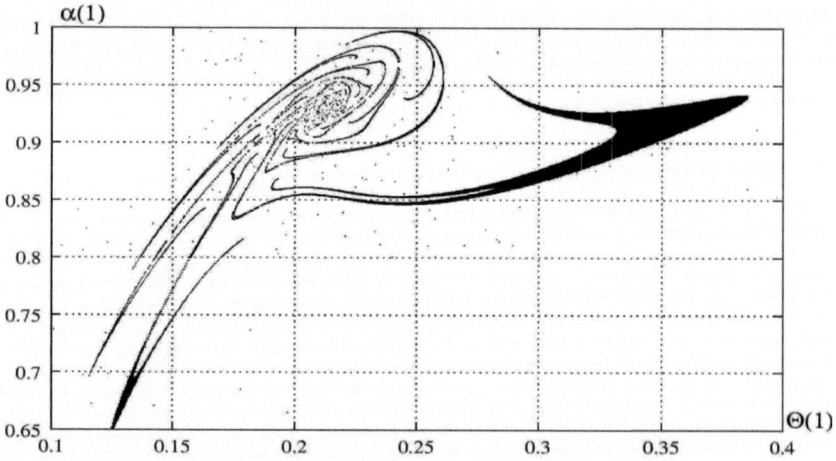

Fig. 2.4. *Total attraction area of the hexa-periodic orbit*

When a part of the area is scaled up (Fig.2.5), its fractal structure is revealed.

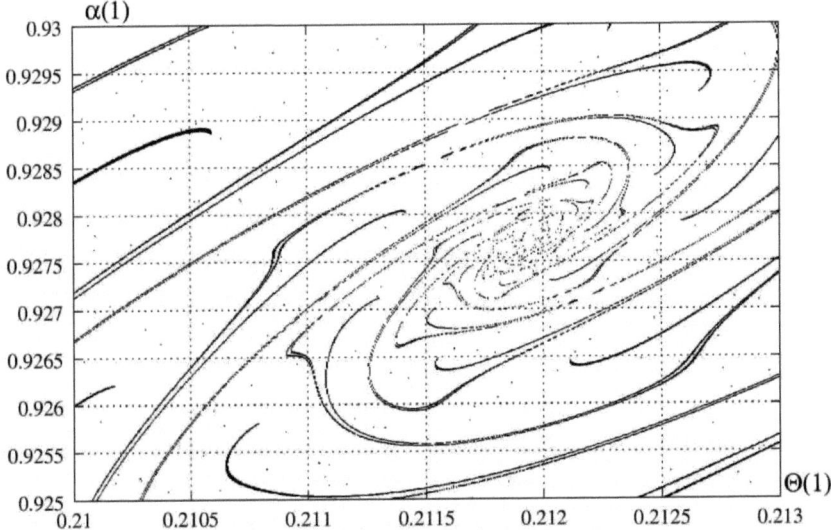

Fig.2.5. *Fragment of the area marked in Fig.2.4*

To obtain an explicit verification of this, the topological dimensions of the discussed structure were determined from the box method in accordance with the following equation (Peitgen, Jürgens & Saupe, 1992; Smuła, Russo & Continillo, 2005):

$$D = -\frac{\lg(N) - \lg(N_*)}{\lg(s) - \lg(s_*)} \qquad (2.6)$$

which, for relatively small value of s, is reduced to:

$$D = -\frac{\lg(N)}{\lg(s)}. \qquad (2.7)$$

In the above equations s is the so called *distribution ratio*. It is the inverse of the number of intervals into which particular sides of the examined area were divided. However, N is the number of boxes comprising the fragments of the discussed geometric structure. s and s_* as well as N and N_* are arbitrary, they differ in the distribution ratio values and in number of boxes. Equations (2.6) and (2.7) are defining expressions of determining dimensions by means of the box method. To employ equation (2.7), the value of s should theoretically tend to zero. Thus, in the numerical analysis it should be set as low as possible. In practice, however, it cannot be lower than the shortest distance between two points of the investigated structure (for example: a strange attractor), as it is not a continuous line. The assumption of a lower value of s is pointless and leads to unreliable results. As far as equation (2.6) is concerned, the values of s and s_* are not that important, and may be assumed arbitrarily. Dimension D is the inclination angle tangent of straight line (Peitgen, Jürgens & Saupe, 1992). The dimension of the structure shown in Fig.2.4 is equal to the value of $D= 1.42$, whereas the dimension of the structure shown in Fig.2.5 equal to $D=1.39$. As the above dimensions are practically the same, it may be concluded that the area of attraction of the hexa-periodic orbit is fractal. Furthermore, the value of the topological dimension of the strange reactor (big dots in Fig.2.3) was derived as $D= 1.38$. Each of the geometric structures mentioned above is characterized by a bigger or smaller degree of dissipation on the phase surface. Such dissipation results from the sensitivity of the system to the changes in temperature and intensity of the reacting mixture at the initial moment. Bigger dissipation denotes greater sensitivity, which, in turn, practically means lower predictability. This means that in practice, even a small change in the initial values or temperature or concentration leads to severe changes in the values of these parameters in the stable state or to changes in the nature of the solution of the model reactor. The rate of the dissipation is the entropy of the system, which may be expressed as:

$$E = -N(Ns^2)\lg_2(Ns^2) \qquad (2.8)$$

where (Ns^2) is the probability of finding the investigated structure on the phase surface. Equation (2.8) is a defining expression of entropy, the general form of which is: $E = -\sum_i p_i \lg_2(p_i)$ (Baker & Gollub, 1996), where p_i is the probability of incidence.

In the case discussed in the paper, the number of all boxes is $M = 1/s^2$. The number of boxes containing the investigated structure is N. Thus, the probability of the incidence of the point of the structure on limited surface is $p_i = N/M = Ns^2 = const$. By combining equations (2.7) and (2.8), the relation between fractal dimension D and entropy E may easily be proved. It assumes the following form:

$$E = (D-2)s^{2(1-D)}\lg_2(s).\qquad\qquad(2.9)$$

Equations (2.8) and (2.9) indicate explicitly that the dissipation disappears in two extreme cases. The first one occurs when there is no geometric structure on the phase surface ($N = 0$). The other one occurs when the structure covers the surface in 100 % ($N = 1/s^2$, $D = 2$). In both cases, entropy $E=0$, which, from the physical point of view, is evident. Accordingly, there must be a certain maximum of entropy within the range from $N=0$ to $N=1/s^2$. On the grounds of equation (2.8) it may be proved that this maximum occurs for:

$$Ns^2 = \frac{1}{\sqrt{e}}\qquad\qquad(2.10)$$

and equals to:

$$E_{max} = \frac{1}{s^2 e \ln(4)}.\qquad\qquad(2.11)$$

In the cases examined within the scope of this study the entropy contained in the structure shown in Fig.2.3 has the value of $E=1891.7$ – which, while assuming that $s=0.001$, constitutes 0.711 % of the maximal entropy; whereas the entropy contained in the structure shown in Fig.2.4 has the value of $E=1290.8$, which constitutes 0.425 % of the maximal entropy; whereas the entropy contained in the strange attractor (big dots in Fig.2.3) has the value of $E=1129.5$ – which constitutes 0.425 % of the maximal entropy.

However, the absolute values of the above parameters are not as essential as their mutual relationships. Thus, assuming the entropy of the strange attractor as the basis, the entropy contained in the structure shown in Fig. 2.4 is found to be 1.67 times bigger than the entropy contained in the strange attractor; whereas the entropy contained in the structure shown in Fig. 2.5 is found to be 1.14 times bigger than the entropy contained in the strange attractor. Hence it may be stated that within the basin boundary the transfer of the reactor from the chaotic state to the stable hexa-periodic oscillation state is less predictable than the change taking place in the chaotic graph of the temperature and flux concentration.

In the calculations derived in the paper the following values of the parameters were assumed: $Da=0.15$, $\gamma = 15$, $n = 1.5$, $\beta = 2$, $f = 0.7$, $\Theta_H = -0.02875$, $\delta = 3$.

References

Baker L.B. & Gollub J.P. (1996). Chaotic dynamics: an introduction. *Cambridge University Press.*

Berezowski M. (2000). Spatio-temporal chaos in tubular chemical reactors with the recycle of mass. *Chaos, Solitons & Fractals, 11,* 1197-1204.

Berezowski M. (2001). Efect on delay time on the generation of chaos in continuous systems. One-dimensional model. Two-dimensional model – tubular chemical reactor with recycle. *Chaos, Solitons & Fractals, 12,* 83-89.

Berezowski M. (2006). *Fractal character of basin boundaries in a tubular chemical reactor with mass recycle.* Chemical Engineering Science, 61/4, 1342-1345, 2006.

Jacobsen E.W. & Berezowski M. (1998). Chaotic dynamics in homogeneous tubular reactors with recycle. *Chemical Engineering Science, 53,* 4023 – 4029.

Luss D. & Amundson N.R. (1967). Stability of loop reactors. *A.I.Ch.E. J., 13,* 279-290. Nusse H.E. & Yorke J.A. (1994). Dynamics: numerical explaration. Springer-Verlag New York, Inc.

Peitgen. H. –O., Jürgens H. & Saupe D. (1992). Fractals for the Classroom. Part 1. *Springer-Verlag* New York.

Reilly M.J. & Schmitz R.A. (1966). Dynamics of a tubular reactor with recycle. Part I. *A.I.Ch.E. J., 12,* 153-161.

Reilly M.J. & Schmitz R.A. (1967). Dynamics of a tubular reactor with recycle. Part II. *A.I.Ch.E. J., 13,* 519-527.

Smuła J., Russo L. & Continillo G. (2005). Computation of fractal dimension for cascade of three CSTR with periodic feed switching. *The seventh Italian Conference on Chemical & Process Engineering.*

Chapter 3. Liapunov's time

The dynamics of a homogeneous tubular chemical reactor without dispersion and with external feedback was discussed, for example, in [1–8], where it was demonstrated that the concentration and temperature of the fluid flux may be constant in the steady state or may oscillate in the form of rectangular time wave. The oscillations may be more or less complex, depending on the periodicity of the time series of the above mentioned variables of a given state. The oscillations may also have a chaotic nature. In this chaper the theoretical analysis entailed a mathematical model of a homogeneous tubular chemical reactor without dispersion and with external recirculation of mass described in [1,2,4–6,8]. The main objective of the research was the determination of the values of the so called Liapunov's time, after which the prediction of the chaotic solutions of the model becomes practically impossible, as the accuracy of designating the state of the system equals the size of the entire attractor [5]. Liapunov's time certainly depends on the initial accuracy of the state of the system: the bigger the accuracy, the longer Liapunov's time. For the initial accuracy approaching infinity (that is: the initial distance between the two trajectories is approaching zero), Liapunov's time also approaches infinity. This means that the entire course of the chaotic trajectory is predictable. In practice, however, the accuracy of the initial state of a dynamic system is always limited. In physics the obstacle is Planck's constant, whereas in numerical calculations it is the length of the machine word. Experience shows that even for very big accuracy of the initial state of the system, the value of Liapunov's time is relatively low. A numerical and differential analysis was conducted on a mathematical model of a tubular chemical reactor with mass recycle [5]:

$$\frac{d\alpha_{k+1}(\xi)}{d\xi} = \phi[\alpha_{k+1}(\xi), \Theta_{k+1}(\xi)] \tag{3.1}$$

$$\frac{d\Theta_{k+1}(\xi)}{d\xi} = \phi[\alpha_{k+1}(\xi), \Theta_{k+1}(\xi)] + \delta[\Theta_H - \Theta_{k+1}(\xi)] \tag{3.2}$$

$$\alpha_{k+1}(0) = f\alpha_k(1); \quad \Theta_{k+1}(0) = f\Theta_k(1) \tag{3.3}$$

$$\phi(\alpha, \Theta) = Da(1-\alpha)^n \exp\left(\gamma \frac{\beta\Theta}{1+\beta\Theta}\right). \tag{3.4}$$

The objective was to designate the impact of the dimensionless temperature of the cooling agent Θ_H on Liapunov's time

$$T = \frac{1}{\lambda_+} \ln \frac{d}{\varepsilon} \qquad (3.5)$$

after which any prediction of the solutions for the model of the reactor is impossible. In the above equation, Liapunov's exponent $\lambda_+ > 0$ was derived from the following relation:

$$\lambda_+ = \lim_{k \to \infty} \frac{1}{k} \ln \max|s_1, s_2| \qquad (3.6)$$

In the designation of time T only positive values of the exponent were considered. For $\lambda_+ \leq 0$ the solutions of the model are periodic, which means that $T = \infty$. Variables s_1 and s_2 are eigenvalues of the following matrixes [5]:

$$f^k \prod_{j=1}^{k} \exp\left(\int_0^1 \overline{J}_j d\xi\right) \qquad (3.7)$$

whereas $\overline{J}$ is Jacobie's matrix derived from the right sides of Eqs. (3.1) and (3.2).

It was assumed in the calculations that the initial deviation of the trajectories (initial accuracy of the state of the system) is equal to $\varepsilon = 10^{-10}$. Their maximal deviation, which is diameter d of the chaotic zone, was determined in a numerical way from Henon's attractor. This diameter is, at the same time, the minimal accuracy with which the dynamic state of the system can be designated. For example, on the grounds of Feigenbaum's diagram discussed in [1] and assuming that $Da=0.15$, $n=1.5$, $\gamma=15$, $\beta=2$, $f=0.7$, $\delta=3$, $\Theta_H=-0.02878$, Henon's attractor was designated in a manner identical as in Fig. 3.1.

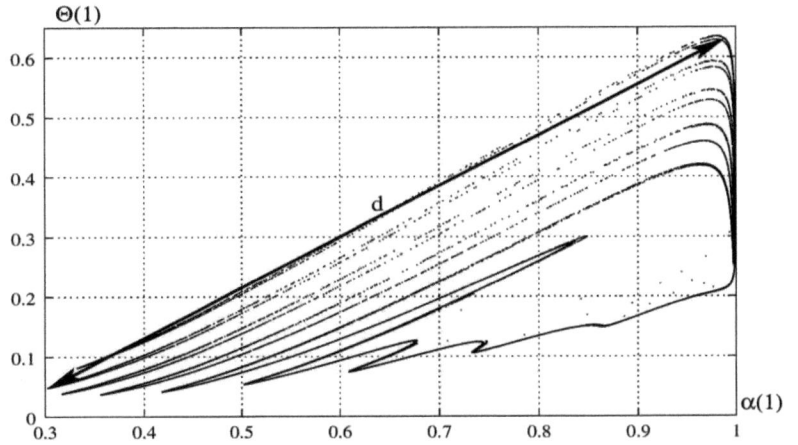

Fig. 3.1. *Henon's attractor of the reactor model.* $\Theta_H = -0.02878$

In this example, the diameter has the value of $d = 0.89$. By means of the same method, the graph of the changes of diameter d was plotted as a function of temperature Θ_H – see Fig.3.2.

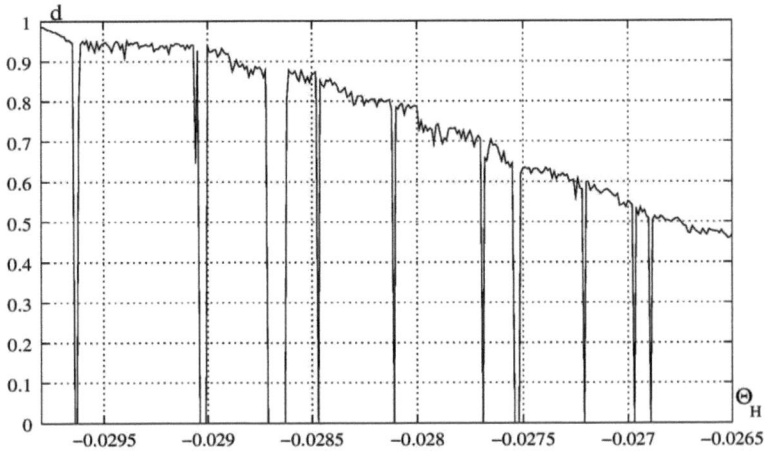

Fig. 3.2. *Nature of changes in diameter d as function* Θ_H

The graph has a fractal nature. The value of $d = 0$ confirms a periodic solution of the model, whereas $d > 0$ confirms a chaotic solution. On the bases of the above mentioned results the graph of Liapunov's time T was plotted as a function of temperature Θ_H, as shown in Fig.3.3.

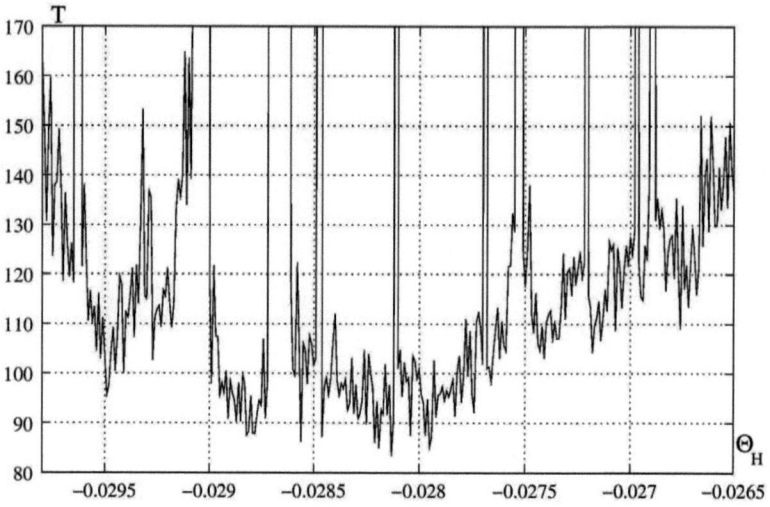

Fig. 3.3. *Nature of changes in Liapunov's time T as function* Θ_H

The graph also has a fractal nature. The value of $T \to \infty$ confirms a periodic solution of the model, whereas $T < \infty$ a chaotic one. It may be concluded from Fig. 3.3 that, if, for example, $\Theta_H = -0.02878$ is assumed, the value of Liapunov's time for the tested model is equal to $T=88$. This result was confirmed on the time series in Fig. 3.4, from which it may be deduced that starting with $\tau = 88$ the two time graphs begin to differ from each other rather significantly.

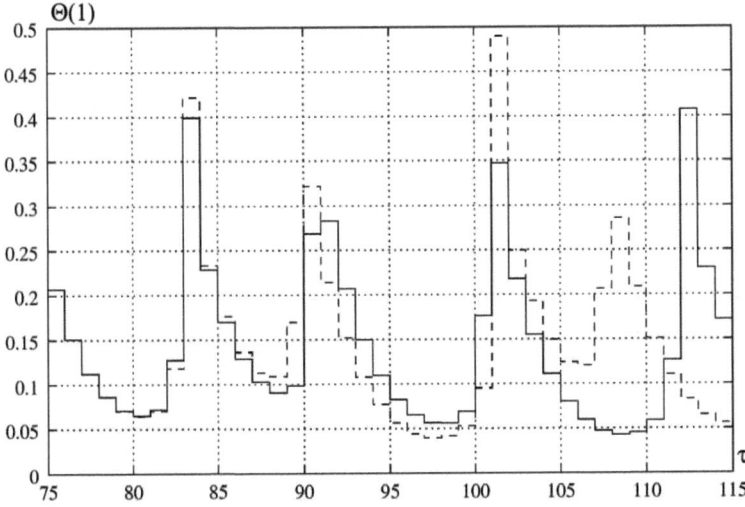

Fig. 3.4. *The time series - sensitivity to the change in initial conditions.*

$$\Theta_H = -0.02878$$

References

[1] Berezowski M. *Spatio-temporal chaos in tubular chemical reactors with the recycle of mass.* Chaos, Solitons & Fractals 2000;11:1197–204.

[2] Berezowski M. *Effect of delay time on the generation of chaos in continuous systems. One-dimensional model. Two-dimensional model – tubular chemical reactor with recycle.* Chaos, Solitons & Fractals 2001;12:83–9.

[3] Berezowski M, Grabski A. *Chaotic nad non-chaotic mixed oscillations in a logistic system with delay and heat-integrated tubular chemical reactor.* Chaos, Solitons & Fractals 2002;14:97–103.

[4] Berezowski M. *Fractal solutions of recirculation tubular chemical reactors.* Chaos, Solitons & Fractals 2003;16:1–12.

[5]. Berezowski M. *Liapunov's time of a tubular chemical reactor with mass recycle.* Chaos, Solitons&Fractals, **41**/5, 2647-2651, 2009.

Chapter 4. Parametric continuation method

Mathematical models that describe equipment and systems used in chemical engineering often offer ambiguous solutions, due to the so called: "multiple steady states phenomenon", where specifically determined parameters values correspond to more than one variable values such as: temperature, concentration, pressure, etc. (Berezowski, 1987; Berezowski & Burghardt, 1989; Burghardt & Berezowski, 1990; Subramanian & Balakotaiah, 1996). The steady state diagrams created on the bases of this may have a very complicated nature, presenting a tangle of lines on the plane or in space (Berezowski, 2000). In view of a non-linear character of the models, often described by means of differential equations, their analytical solution is impossible, thus, leaving space for numerical methods. Nevertheless, also the analytical methods entail essential calculation problems arising from the ambiguity of solutions and small distances between specific steady states (Berezowski, 2000- Fig. 3). In such cases, iteration methods that render solutions dependent on the initial values are practically useless. Therefore, the parametric continuation method, which is not based on numerical iteration, may be helpful, especially that it is not sensitive to the degree of complexity of the graph (Seydel, 1994; Shalashin and Kuznetsov, 2003). In this chapter simple ways of the application of the parametric continuation method are discussed on the bases of chemical reactor models (Berezowski, 2010). The first example involves a non-adiabatic tank reactor described by simple differential equations, whereas the second one concerns adiabatic tubular reactor with longitudinal dispersion, described by simple differential equations of the second order with boundary conditions. Unlike the manner based on the so called: " fictitious dynamics" (Berezowski, 2000) the discussed method does not require the installation of huge professional IT systems (Doedel et al., 1997).

Let us assume an n- dimensional model:

$$\overline{F}(\overline{y}, p) = \overline{0} \qquad (4.1)$$

from which n characteristics (diagrams) should be designated as $y_i(p)$ in a specified interval of the variability of parameter p. If the non-linear equation system (4.1) is mathematically confounding, the analytical designation of particular diagrams seems to be impossible. In such case, it is safe to fall back on numerical methods, one of which could involve an iteration solution of the system for the successive values of p, for example, on the bases of Newton's method. Yet, the main problem arises when the solutions of system (4.1) are ambiguous, i.e. when a given value of p corresponds to more than one vector $\bar{y}(p)$. In such case, the sought solutions are very sensitive to changes in the initial values, especially when the solutions are located in very close proximity. In the majority of cases, it is practically impossible to derive correct diagrams. Thus, other methods have to be employed, for example, the parametric continuation method, the description and application of which is presented below.

By designating the total differential of system (1) we obtain:

$$\bar{\bar{J}}d\bar{y} + \bar{w}dp = \bar{0} \qquad (4.2)$$

where $\bar{\bar{J}}$ is Jacobi matrix in the following form:

$$\bar{\bar{J}} = \left\{\frac{\partial f_i}{\partial y_j}\right\}; \ i=1,n \qquad (4.3)$$

whereas, $\bar{w}$ is the partial derivative vector calculated in terms of parameter p:

$$\bar{w} = \left\{\frac{\partial f_i}{\partial p}\right\}; \ i=1,n. \qquad (4.4)$$

From equation (2) owe derive that:

$$d\bar{y} = -\bar{\bar{J}}^{-1}\bar{w}dp. \qquad (4.5)$$

If the above dependence is to be used in numerical calculations, differential increases $d\bar{y}$ and dp should be substituted by the following difference increases:

$$\Delta\bar{y} = \bar{y}_{K+1} - \bar{y}_k, \ \Delta p = p_{k+1} - p_k \qquad (4.6)$$

which leads to the following relation:

$$\bar{y}_{k+1} = \bar{y}_k - \bar{\bar{J}}_k^{-1}\bar{w}_k\Delta p. \qquad (4.7)$$

If the computational process is started with precisely determined values of $\bar{y}_0$ and p_0 (that is, fulfilling equation (4.1) with big accuracy), the successive values of vector $\bar{y}_k$, determining the next points of the diagrams, are designated without the necessity of using any other iterations. However, the discussed method does not eliminate the problem involved in the ambiguity of the solutions of equations system (4.1). As shown in conceptual scheme 4.1, at bifurcation limit points LP there is a change of the direction of designating parameter p.

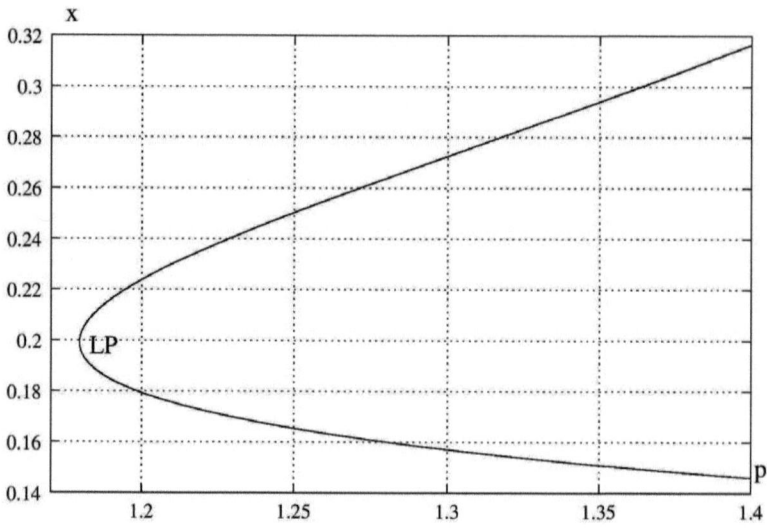

Fig.4.1. *Conceptual diagram*

Thus, the points should be determined numerically in order to change the sign of increase. Δp. It should be pointed out that at points *LP*, due to the occurrence of the extreme value of parameter p in relation to variable y, differential increase $dp = 0$ and matrix equation (4.2) is transformed as follows:

$$\overline{\overline{J}}d\overline{y} = \overline{0}.\tag{4.8}$$

The above equation is met only when matrix determinant $\overline{\overline{J}}$ is equal to zero:

$$\det \overline{\overline{J}} = 0.\tag{4.9}$$

This means that, the determinant changes its sign at points *LP*. Hence, specific values of parameter p should be derived from the following relation:

$$p_{k+1} = p_k + sign(\det \overline{\overline{J}}_k)\Delta p.\tag{4.10}$$

At the same time, it should be emphasized that in view of the fact that the determinant zeros points *LP*, matrix $\overline{\overline{J}}$ becomes peculiar, which means that its inverse form cannot be derived. However, in view of this, there is a limit:

$$| \lim_{\Delta p \to 0} \frac{\Delta p}{\det \overline{J}} |_{LP} < \infty \qquad (4.11)$$

The computational process may be continued at points LP.

The method discussed above was exemplified on a tank reactor and an adiabatic reactor with longitudinal dispersion.

Dimensionless model of a non-adiabatic tank reactor.

The mass balance:

$$\frac{d\alpha}{d\tau} + \alpha = \phi(\alpha, \Theta) \qquad (4.12)$$

the heat balance:

$$Le \frac{d\Theta}{d\tau} + \Theta = \phi(\alpha, \Theta) + \delta(\Theta_c - \Theta) \qquad (4.13)$$

the reaction kinetics function:

$$\phi(\alpha, \Theta) = Da(1 - \alpha)^n \exp\left(\gamma \frac{\beta\Theta}{1 + \beta\Theta} \right). \qquad (4.14)$$

According to the definition of the steady states, they may be designated by equating both time derivatives in the above equations to zero. Accordingly, we obtain the following system of equations:

$$f_1 = -\alpha + \phi(\alpha, \Theta) = 0 \qquad (4.15)$$
$$f_2 = -\Theta + \phi(\alpha, \Theta) + \delta(\Theta_c - \Theta) = 0. \qquad (4.16)$$

from which $\alpha(p)$ and $\Theta(p)$ steady state characteristics may be derived for any model parameter. In such case, the elements of Jacobi matrix assume the following form:

$$j_{11} = -1 - nDa(1 - \alpha)^{n-1} \exp\left(\gamma \frac{\beta\Theta}{1 + \beta\Theta} \right) \qquad (4.17)$$

$$j_{12} = Da(1 - \alpha)^n \exp\left(\gamma \frac{\beta\Theta}{1 + \beta\Theta} \right) \frac{\gamma\beta}{(1 + \beta\Theta)^2} \qquad (4.18)$$

$$j_{21} = -nDa(1 - \alpha)^{n-1} \exp\left(\gamma \frac{\beta\Theta}{1 + \beta\Theta} \right) \qquad (4.19)$$

$$j_{22} = -1 + Da(1 - \alpha)^n \exp\left(\gamma \frac{\beta\Theta}{1 + \beta\Theta} \right) \frac{\gamma\beta}{(1 + \beta\Theta)^2} - \delta. \qquad (4.20)$$

If, for example, we assume that the selected bifurcation parameter is n (order of reaction), the elements of vector $\overline{w}$ will have the following form:

$$w_1 = w_2 = Da(1-\alpha)^n \exp\left(\gamma \frac{\beta\Theta}{1+\beta\Theta}\right) \ln(1-\alpha). \qquad (4.21)$$

At the same time, it is worth mentioning that if elements j_{21} and j_{22} of matrix $\bar{\bar{J}}$ are divided by Lewis number Le, the eigenvalues of the matrix transformed in such manner indicate the nature of the dynamical solutions of the model. Thus, if the eigenvalues of the matrix are imaginary, we are dealing with Hopf bifurcation HB, determining oscillations (Jacobsen & Berezowski, 1998). So, after considering equations (4.17) - (4.21) in (4.7) and (4.10), a diagram of the tested model was derived – see Fig. 4.2.

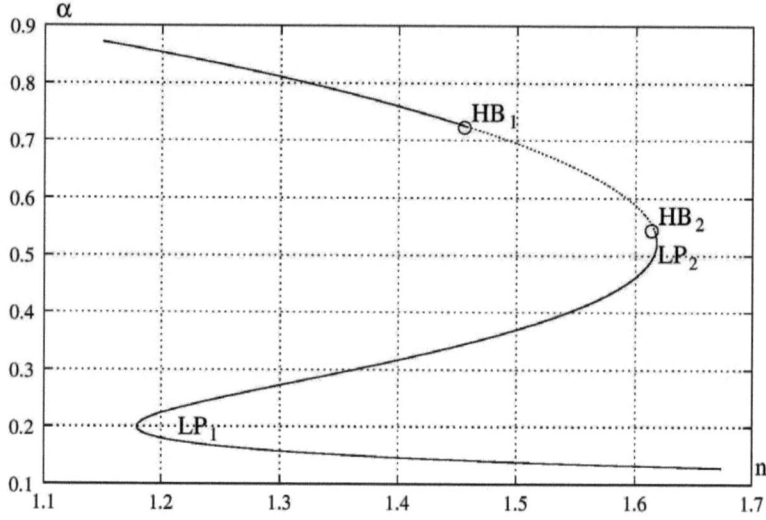

Fig.4.2. *Bifurcation diagram of a tank reactor*

The following parameter values were assumed in the calculations: $Le=1.5$, $Da=0.2$, $\gamma = 20$, $\beta = 1$, $\delta = 2$, $\Theta_C = -0.08$. Bifurcation points LP visible in the diagram determine changes in the number of the steady states, whereas points HB determine the generation of oscillations.

Dimensionless model of an adiabatic tubular reactor with longitudinal dispersion.

The balance equations concerning the steady state have the following form:

the mass balance:

$$\frac{d\alpha}{d\xi} = \frac{1}{Pe_M}\frac{d^2\alpha}{d\xi^2} + \psi(\alpha, \Theta) \qquad (4.22)$$

the heat balance:

$$\frac{d\Theta}{d\xi} = \frac{1}{Pe_H}\frac{d^2\Theta}{d\xi^2} + \psi(\alpha, \Theta). \qquad (4.23)$$

The boundary conditions ascribed to the above system of equations are as follows:

$$\alpha(0) = \frac{1}{Pe_M}\frac{d\alpha}{d\xi}\Big|_{\xi=0} \quad ; \quad \Theta(0) = \frac{1}{Pe_H}\frac{d\Theta}{d\xi}\Big|_{\xi=0} \qquad (4.24)$$

$$\frac{d\alpha}{d\xi}\Big|_{\xi=1} = 0 \; ; \; \frac{d\Theta}{d\xi}\Big|_{\xi=1} = 0. \qquad (4.25)$$

Assuming that $Pe_M = Pe_H = Pe$, it is easy to demonstrate that relation $\Theta = \alpha$ holds in the steady state. Under such circumstancesthe following system of equations is reduced to the singular equation:

$$\frac{d\alpha}{d\xi} = \frac{1}{Pe}\frac{d^2\alpha}{d\xi^2} + \phi(\alpha) \qquad (4.26)$$

with the following boundary conditions:

$$\alpha(0) = \frac{1}{Pe}\frac{d\alpha}{d\xi}\Big|_{\xi=0} \quad ; \quad \frac{d\alpha}{d\xi}\Big|_{\xi=1} = 0 \qquad (4.27)$$

where the kinetics reation function is:

$$\phi(\alpha) = Da(1-\alpha)^n \exp\left(\gamma\frac{\beta\alpha}{1+\beta\alpha}\right). \qquad (4.28)$$

By inserting auxiliary variable:

$$u = \frac{d\alpha}{d\xi} \qquad (4.29)$$

the system of equations (26) – (27) is transformed to the form:

$$\frac{du}{d\xi} = Pe(u - \phi(\alpha)) \qquad (4.30)$$

with boundary conditions:

$$\alpha(0) = \frac{1}{Pe} u(0); \quad u(1) = 0. \tag{4.31}$$

In the course of the analysis of the above system of equations it may be proved that the integration of equations (4.29) and (4.30) within the range from $\xi = 0$ to $\xi = 1$ is unstable. Thus, their numerical solution is practically impossible. Accordingly, inverse integration, i.e. within the range from $\xi = 1$ to $\xi = 0$ should be employed. To designate a complete diagram of the steady states the following function should be defined:

$$f(\alpha(1)) = \alpha(0) - \frac{1}{Pe} u(0) = 0 \tag{4.32}$$

which means, that the discussed problem is reduced to one-dimensional form. So, under such circumstances equations (7) and (10) are transformed into the following form:

$$\alpha_{k+1}(1) = \alpha_k(1) - \frac{\frac{\partial f}{\partial p}_{|k}}{\frac{\partial f}{\partial \alpha(1)}_{|k}} \Delta p \tag{4.33}$$

$$p_{k+1} = p_k + sign\left(\frac{\partial f}{\partial \alpha(1)}_{|k}\right)\Delta p. \tag{4.34}$$

The partial derivatives in the above equations should be designated numerically, which is not difficult. Assuming Damköhler number Da as the bifurcation parameter, the diagram of the steady states shown in Fig.4.3 was derived. Other assumed parameters values were: $\gamma = 15$, $\beta = 2$, $n = 1,5$, $Pe = 100$.

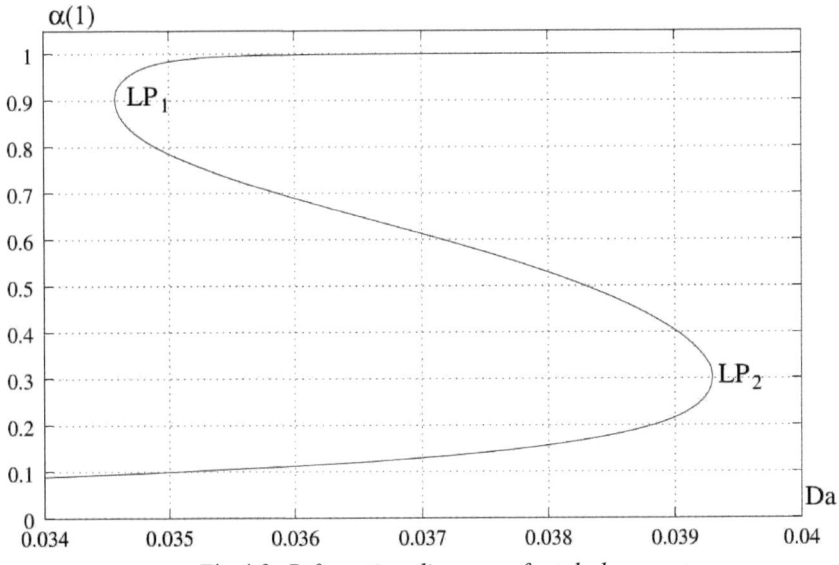

Fig.4.3. *Bifurcation diagram of a tubular reactor*

References

Berezowski, M. (1987). Multiple steady states in adiabatic tubular reactors with recycle. *Chemical Engineering Science, 52,* 1207-1210.

Berezowski, M. & Burghardt, A. (1989). A generalized analytical method for determination multiplicity features in chemical reactors with recycle. *Chemical Engineering Science, 44,* 2927-2933.

Burghardt, A. & Berezowski, M. (1990). Analysis of the structure of steady-state solutions for porous catalytic pellets-first-order reversible reactions. *Chemical Engineering Science, 45,* 705-719.

Berezowski, M. (2000). Method of determination of steady-state diagrams of chemical reactors. *Chemical Engineering Science, 55,* 4291-4295.

Berezowski, M. (2010). *The application of the parametric continuation method for determining steady state diagrams in chemical engineering.* Chemical Engineering Science , 65/19, 5411-5414, 2010.

Doedel, E. J., Champneys, A. R., Fairgrevie, T. F., Kuznetsov, Y. A., Sandstede, B. & Wang, X. (1997). AUTO97: Continuation and bifurcation software for ordinary differential equations. Technical Report, Computational Mathematics Laboratory, Concordia Univesrity. Also http://indy.cs.concordia.ca/auto.

Jacobsen, E. W. & Berezowski, M. (1998). Chaotic dynamics in homogeneous tubular reactors with recycle. *Chemical Engineering Science, 53*, 4023-4029.

Seydel, R. (1994). Practical bifurcation and stability analysis. Springer, Berlin.

Shalashin V.I. & Kuznetsov E.B. (2003). Parametric continuation and optimal parametrization in applied mathematics and mechanics. Kluwer Academic Publisher, AA Dordrecht, The Netherlands.

Subramanian, S. & Balakotaiah, V. (1996). Classification of steady state and dynamic behavior of distributed reactor models. *Chemical Engineering and Processing, 51*, 401-421.

Chapter 5. Bifurcation analysis and relaxation method

Industrial processes that involve chemical reactions often use continuous stirred tank reactors (CSTR). The basic task for designers of reactor systems is to select apparatuses and their configuration to provide the highest possible conversion degree of inflowing raw material. There are many elaborations and analyses of tubular reactors with reverse flow (Sheintuch, 2005; Sheintuch & Nekhamkina, 2004; Glöckler, Kolios & Eigenberger, 2003; Jeong & Luss, 2003; Annalad, Scholts, Kuipers, & Swaaij, Part I and Part II, 2002; Khinast, Jeong & Luss, 1999; Řeháček, Kubiček & Marek, 1998; Purwono, Budman, Hudgins, Silveston & Matros, 1994; Gupta & Bhata, 1991). The analysis of the dynamics of reverse flow system used for cascade tank reactors was discussed in (Żukowski & Berezowski, 2000; Kulik & Berezowski, 2008; Berezowski & Kulik, 2009), where a possibility of the ocurrence of chaotic oscillations was proved.

According to industrial and laboratory practice, as well as to theoretical analyses, for certain parameter ranges, the cascade enables good conversion of the raw material, whereas, for other ranges very low conversion degrees are achieved, meaning that practically chemical reactions do not occur. One of the ways of increasing the conversion degree in such systems is to use cyclical reverse-flow of the raw material. Such method makes it possible to store the generated heat energy. For tubular gas reactors with fixed-bed this energy is accumulated, first and foremost, in the catalyst, whereas, in the case of liquid tank reactors the heat is collected in the reaction mix.

Likewise, it is evident that in the reverse flow system, high conversion degree may only be obtained for specific values of the system parameters, including the time between successive changes of the flow direction. The method presented in this chaper enables an increment of the conversion degree also when the reverse flow renders low conversion (Berezowski, 2011). It involves - after the change in the direction of the feed flux - cutting off, for a certain time, the flow to the reactors, consisting of a low-concentration product mix and the simultaneous supply and receipt of the product from the apparatuses containing low concentrations of the raw material. Such time is referred to as relaxation time. Thanks to this procedure the raw material in the cut-off reactors has better chance of higher conversion. After the passage of the relaxation time, the previously cut-off reactors are again fed with fluxes and the substance subjected to conversion is collected. The theoretical analysis of the above-mentioned systems was conducted on the bases of applicable bifurcation diagrams and time series. They concern both steady states, as well as enforced oscillation states, which are a natural consequence of reverse-flow applications. The method of designating bifurcation diagrams was discussed in (Berezowski, 2010). The analyzed cascade consists of two identical adiabatic continuous stirred tank reactors CSTR 1 and CSTR 2 (Fig.5.1).

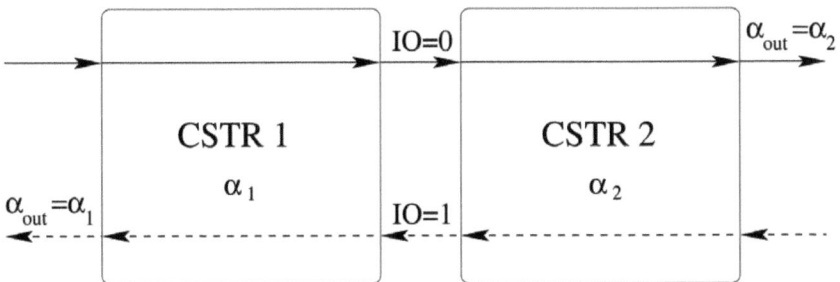

Fig. 5.1. *Schematic diagram of the cascade of continuous stirred tank reactors with constant and reverse flow direction*

The balance equations for particular apparatuses are expressed by the following differential system:

$$\frac{d\alpha_1}{d\tau} + \alpha_1 = IO\alpha_2 + \phi(\alpha_1) \tag{5.1}$$

$$\frac{d\alpha_2}{d\tau} + \alpha_2 = (1 - IO)\alpha_1 + \phi(\alpha_2) \tag{5.2}$$

where IO is a variable determining the direction of the reacting flux flow. Accordingly, if the flow passes from CSTR 1 to CSTR 2, $IO = 0$. Otherwise, $IO = 1$. The use of this variable will be very important in the reverse-flow system applied to the discussed cascade. It may be inferred from the adiabatic nature of the apparatusses that the temperature of the reactive mass is proportional to its concentration, which, in the assumed dimensionless notations means that $\Theta = \alpha$. Hence, it is unnecessary to introduce additional equations referring to the reacting heat into the model. Whitin the scope of the chapter, the bifurcation diagram of the cascade was determied at first, assuming that the feed flux does not change its direction. It may be inferred from the adiabatic nature of the reactors that the occurrence of autonomuos temperature and concentration oscillations is impossible. This means that the only attractors may be the steady states. On the grounds of the parametric continuation method described in (Berezowski, 2010), a diagram of the states was designated and marked with the "ss" line (see Figure 5.2).

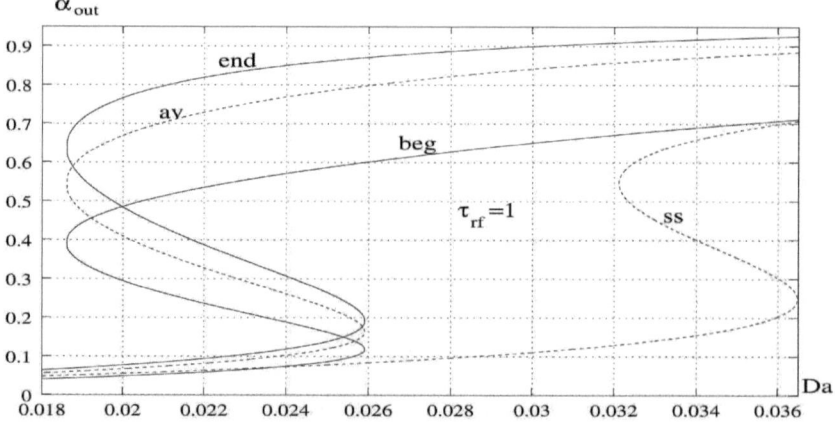

Fig. 5.2. *Conversion degree at the cascade outlet.* $\tau_{rf} = 1$

Variable α_{out} is the conversion degree at the system outlet. In all caculations within the framework of the paper the following parameter vaules were used: $\gamma = 15$, $\beta = 0.65$, $n = 1.5$.

In view of the symmetry of the cascade, the calculations may be based on the assumption of any value of the variable controlling the flow direction, i.e. $IO = 0$ or $IO = 1$. In the first case, $\alpha_{out} = \alpha_2$, whereas, in the second one, $\alpha_{out} = \alpha_1$.

By equations (5.1)-(5.2), it may be concluded that in the steady state the following dependencies apply:

$$f_1 = IO\alpha_2 + \phi(\alpha_1) - \alpha_1 = 0 \qquad (5.3)$$

$$f_2 = (1 - IO)\alpha_1 + \phi(\alpha_2) - \alpha_2 = 0 \qquad (5.4).$$

On the grounds of the parametric continuation method, the following relation is derived:

$$\overline{\alpha}_{k+1} = \overline{\alpha}_k - \overline{\overline{J}}_k^{-1}\overline{w}_k \, sign(\det \overline{\overline{J}}) \Delta p \qquad (5.5)$$

$$p_{k+1} = p_k + sign(\det \overline{\overline{J}}) \Delta p \qquad (5.6)$$

where k is the number of the continuation code, *sign* is the sign of martix $\overline{\overline{J}}$ determinant, Δp an increment of the value of the continuation parameter, whereas:

$$\overline{\alpha} = \begin{bmatrix} \alpha_1 \\ \alpha_2 \end{bmatrix} \qquad (5.7)$$

$$\overline{\overline{J}} = \begin{bmatrix} \dfrac{\partial f_1}{\partial \alpha_1} & \dfrac{\partial f_1}{\partial \alpha_2} \\ \dfrac{\partial f_2}{\partial \alpha_1} & \dfrac{\partial f_2}{\partial \alpha_2} \end{bmatrix} \qquad (5.8)$$

$$\overline{w} = \begin{bmatrix} \dfrac{\partial f_1}{\partial p} \\ \dfrac{\partial f_2}{\partial p} \end{bmatrix} \qquad (5.9).$$

At the same time:

$$\frac{\partial f_1}{\partial \alpha_1} = \frac{\partial \phi}{\partial \alpha_1} - 1 \qquad (5.10)$$

$$\frac{\partial f_1}{\partial \alpha_2} = IO \qquad (5.11)$$

$$\frac{\partial f_2}{\partial \alpha_1} = 1 - IO \qquad (5.12)$$

$$\frac{\partial f_2}{\partial \alpha_2} = \frac{\partial \phi}{\partial \alpha_2} - 1 \qquad (5.13).$$

Assuming that there is a single $A \to B$ reaction of the $n\text{-}th$ order occuring in the reactors, the kinetic functions have the following form:

$$\phi(\alpha_i) = Da(1-\alpha_i)^n \exp\left(\gamma \frac{\beta \alpha_i}{1+\beta \alpha_i}\right); \ i=1,2 \qquad (5.14).$$

This means that the applicable derivatives are described by dependencies:

$$\frac{\partial \phi}{\partial \alpha_i} = \left[Da(1-\alpha_i)^{n-1}\left(-n+(1-\alpha_i)\frac{\gamma\beta}{(1+\beta\alpha_i)^2}\right)\right]\exp\left(\gamma\frac{\beta\alpha_i}{1+\beta\alpha_i}\right); \ i=1,2 \qquad (5.15).$$

Assuming, for example, that bifurcation parameter p is Damköhler number Da, vector $\overline{w}$ components are, respectively:

$$\frac{\partial f_i}{\partial p} = \frac{\partial f_i}{\partial Da} = (1-\alpha_i)^n \exp\left(\gamma\frac{\beta\alpha}{1+\beta\alpha}\right); \ i=1,2 \qquad (5.16).$$

As indicated in Fig.5.2. high conversion degrees are achieved for $Da > 0.032$, but, in this specific range the phenomenon of the multiplicity of steady states occurs.

In this section of the book a theoretical analysis of the cascade with reverse flow is conducted. The dynamics of this process was already discussed in (Żukowski & Berezowski, 2000; Kulik & Berezowski, 2008; Berezowski & Kulik, 2009), where a possibility of the occurrence of chaotic oscilations was proved. Assuming that the time period betweeen successive changes of the flow direction is τ_{rf}, the basic oscilation period of variable α_{out} at the system outlet is also equal to τ_{rf}. The output is CSTR 2 ($IO=0$) and CSTR 1 ($IO=1$), alternately.

.

In the course of the analysis of the system, the next two bifurcation diagrams were designated, referring, this time, to enforced cycles with time τ_{rf} (Fig.5.2). To achieve this, the parametric continuation method described in (Berezowski, 2010) was reapplied. The two curves marked in Fig.5.2. and labelled: "*beg*" and "*end*" refer to the value of α_{out} at the beginning and at the end of the cycle. Curve "*av*" is the average value of α_{out}, calculated in a given cycle.

After considering periodic change of the raw material flux flow direction in model (5.1)-(5.2), the general balance equations have the following discreted form:

$$\frac{d\alpha_{1,j+1}}{d\tau} + \alpha_{1,j+1} = IO\alpha_{2,j+1} + \phi(\alpha_{1,j+1}) \qquad (5.17)$$

$$\frac{d\alpha_{2,j+1}}{d\tau} + \alpha_{2,j+1} = (1 - IO)\alpha_{1,j+1} + \phi(\alpha_{2,j+1}) \qquad (5.18)$$

with boundary conditions:

$$\alpha_{1,j+1}(\tau_{rf}) = \psi(\alpha_{1,j+1}(0), IO\alpha_{2,j+1}(0)) = \psi(\alpha_{1,j}(\tau_{rf}), IO\alpha_{2,j}(\tau_{rf})) \qquad (5.19)$$

$$\alpha_{2,j+1}(\tau_{rf}) = \psi(\alpha_{2,j+1}(0), (1 - IO)\alpha_{1,j+1}(0)) = \psi(\alpha_{2,j}(\tau_{rf}), (1 - IO)\alpha_{1,j}(\tau_{rf})) \qquad (5.20)$$

where j is the number of the cycle, whereas ψ is an integral transformation of equations (5.17) and (5.18).

In accordance with the previous assumptions, variable $IO = 0$ when j is an even number, but $IO = 1$ when j is an odd number. The conversion degree at the outlet of the system is decsribed by the equation:

$$\alpha_{out,j} = IO\alpha_{1,j} + (1 - IO)\alpha_{2,j} \qquad (5.21).$$

If the sought solutions are fixed oscilation points enforced in time τ_{rf}, the following relations hold for each cycle:

$$\alpha_1(\tau_{rf}) = \alpha_2(0) \qquad (5.22)$$

$$\alpha_2(\tau_{rf}) = \alpha_1(0) \qquad (5.23).$$

Hence, the parametric continuation method should be used for the system of equations:

$$F_1 = \alpha_1(\tau_{rf}) - \alpha_2(0) = 0 \qquad (5.24)$$

$$F_2 = \alpha_2(\tau_{rf}) - \alpha_1(0) = 0 \qquad (5.25)$$

which, subsequently, leads to the following recurrence process:

$$\overline{\alpha}_{k+1}\left(\tau_{rf}\right) = \overline{\alpha}_k\left(\tau_{rf}\right) - \overline{\overline{A}}_k^{-1}\overline{b}_k\, sign\!\left(\det\overline{\overline{A}}\right)\!\Delta Da \qquad (5.26)$$

$$Da_{k+1} = Da_k + sign(\det\overline{\overline{J}})\Delta Da \qquad (5.27).$$

where k is the continuation step number, $sign$ is the sign of martix $\overline{\overline{J}}$ determinant, ΔDa is the increment of the value of Damköhler number, while:

$$\overline{\alpha}\left(\tau_{rf}\right) = \begin{bmatrix} \alpha_1\left(\tau_{rf}\right) \\ \alpha_2\left(\tau_{rf}\right) \end{bmatrix} \qquad (5.28)$$

$$\overline{\overline{A}} = \begin{bmatrix} \dfrac{\partial F_1}{\partial \alpha_1(0)} & \dfrac{\partial F_1}{\partial \alpha_2(0)} \\[2mm] \dfrac{\partial F_2}{\partial \alpha_1(0)} & \dfrac{\partial F_2}{\partial \alpha_2(0)} \end{bmatrix} = \begin{bmatrix} \dfrac{\partial \alpha_1\left(\tau_{rf}\right)}{\partial \alpha_1(0)} & -1 \\[2mm] \dfrac{\partial \alpha_2\left(\tau_{rf}\right)}{\partial \alpha_1(0)} - 1 & \dfrac{\partial \alpha_2\left(\tau_{rf}\right)}{\partial \alpha_2(0)} \end{bmatrix} \qquad (5.29)$$

$$\overline{b} = \begin{bmatrix} \dfrac{\partial F_1}{\partial Da} \\[2mm] \dfrac{\partial F_2}{\partial Da} \end{bmatrix} = \begin{bmatrix} \dfrac{\partial \alpha_1\left(\tau_{rf}\right)}{\partial Da} \\[2mm] \dfrac{\partial \alpha_2\left(\tau_{rf}\right)}{\partial Da} \end{bmatrix} \qquad (5.30).$$

The partial derivatives in the above equations may be easily calculated by numerical methods. The designated $\alpha_1\left(\tau_{rf}\right)$ and $\alpha_2\left(\tau_{rf}\right)$ are the initial and final values of conversion degree α_{out} in a given cycle. In Fig.5.2. they were marked as the lines: "*beg*" and "*end*".

In view of process considerations, the average value of the conversion degree at the outlet, calculated as the integral average in a given cycle (line "*av*" in Fig.5.2) is essential. In comparison with the conversion degree in the cascade with constant flow direction, it is obvious that in the case of reverse-flow, the conversion of the system was considerably increased. Range $0.0185 < Da < 0.032$ is of special importance here, because within this range the cascade without the reverse-flow renders only low conversion degrees. On the other hand, the system with the reverse flow renders significantly higher conversion in the same range. Accordingly, for $Da=0.028$ the use of the reverse flow increases the average conversion by almost 800%. It should also be emphasised that in range $0.0185 < Da < 0.026$ the solutions are characterised by the multiplicity of oscilation states. In Fig.5.3. exemplary time series of conversion degree α_{out} for $\tau_{rf} = 1$ and $Da=0.022$ were plotted.

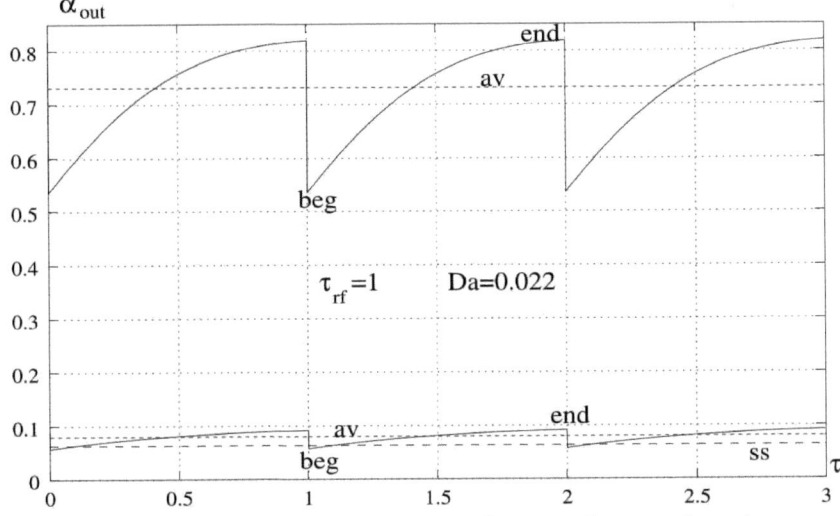

Fig. 5.3. *Time series of the conversion degree at the cascade outlet*

The curves at the top and bottom of the graph refer, respectively, to the upper and lower oscilation state shown in Fig.5.2. Furthermore, curve "*ss*" in Fig.5.3. refers to the conversion degree of the cascade with constant flow direction.

Conversely, in Fig.5.4 the bifurcation diagrams plotted for $\tau_{rf} = 6$ are shown.

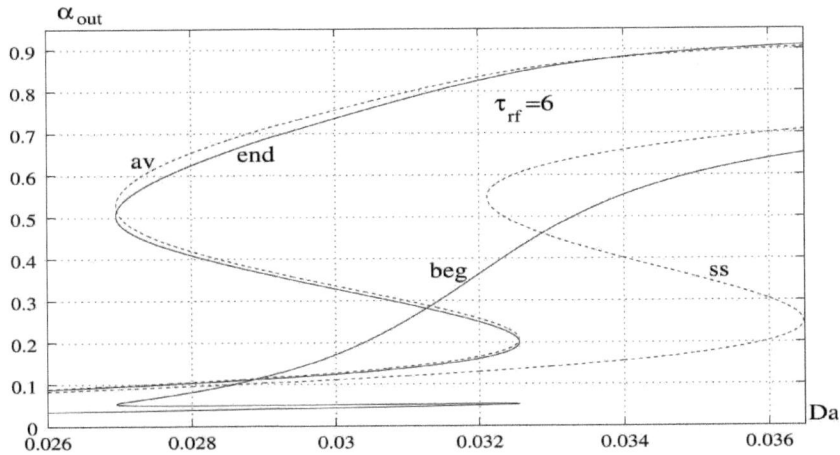

Fig. 5.4. *Bifurcation diagrams of the conversion degree at the cascade outlet.*
$\tau_{rf} = 6$

Also in this case, a significant impact of the reverse flow on the conversion degree is clearly observable. A phenomenon chracteristic for this case is that curve "*av*" is above "*end*" curve. The explanation of this phenomenon is presented by average of the time series in Fig.5.5, where typical extremes are evident.

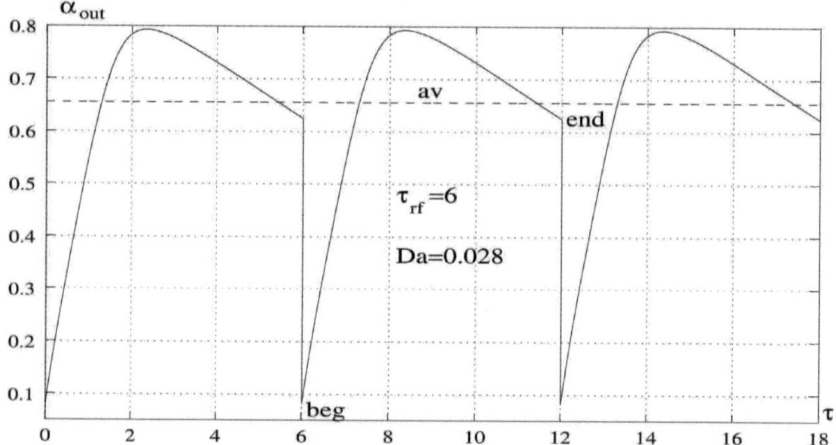

Fig. 5.5. *Time series of the conversion degree at the cascade outlet.* $\tau_{rf} = 6$

As indicated in Fig.5.4, for $Da < 0.027$ the discussed cascade renders only low conversion degrees, despite constant or reverse material flow direction. However, it is possible to significantly increase the average conversion of the raw material. To achieve this, after changing the flow direction, the inflow of the flux to the outlet reactor should be cut off for a certain time, referred to as relaxation time $\tau_{rel} \leq \tau_{rf}$, so that the product is collected only from the reactor into which the raw material is currently fed (Fig.5.6 – case $IO = 1$).

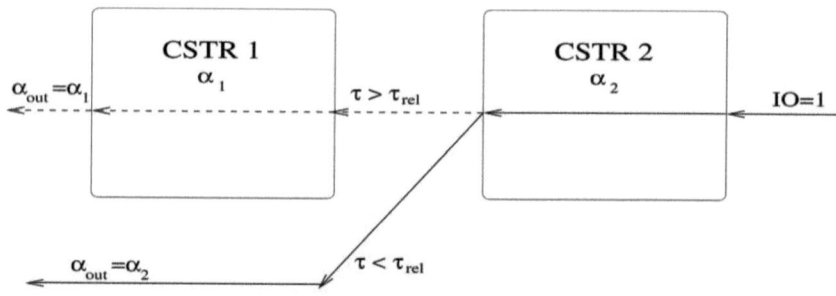

Fig. 5.6. *Schematic diagram of the cascade with relaxation*

Thanks to such procedure, the raw material accumulated in the first reactor will be subjected to conversion, whereas the product amassed in the second reactor will be collected. The value of time τ_{rel} should be selected in a manner that secures the highest average value of variable α_{out}. In Fig.5.7 the time series for variations of α_{out} were plotted for $\tau_p = 6$ and $Da = 0.0265$, and the use of relaxation time $\tau_{rel} = 4.5$, which, for this particular case is the optimal value , i.e. the value that guarantees the highest average values.

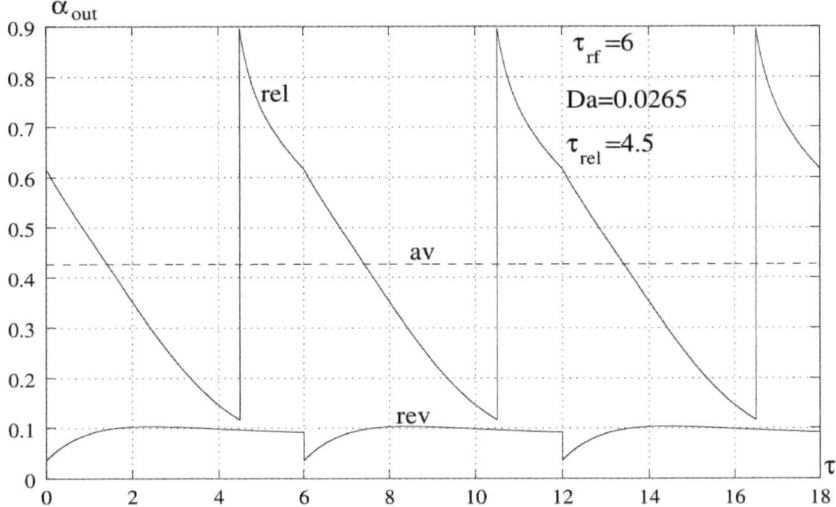

Fig. 5.7. *Time series of the conversion degree at the relaxation cascade outlet.*

$$\tau_{rf} = 6, \ \tau_{rel} = 4.5, \ \tau_{rf} = 6$$

Curve "*rev*" refers to the cascade with the reverse flow, "*rel*" to the relaxation system, whereas line "*av*" is the average value of variable α_{out} after the application of the relaxation procedure. It follows from Fig.5.7. that this average value was increased by about 400 % in comparison with the conversion oferred by the previously discussed systems. It should also be noted that, in some cases, the optimal relaxation time may be equal to the time of cyclical changes in the raw material flow direction, namely: $\tau_{rel} = \tau_p$.

References

Annaland, M., Scholts, H.A.R., Kuipers, J.A.M. & Swaaij, W.P.M. (2002). *A novel reverse flow reactor coupling endothermic and exothermic reactions. Part I: comparison of reactor configurations for irreversible endothermic reactions.* Chemical Engineering Science, 57, 833-854.

Annaland, M., Scholts, H.A.R., Kuipers, J.A.M. & Swaaij, W.P.M. (2002). *A novel reverse flow reactor coupling endothermic and exothermic reactions. Part II: Sequential reactor configuration for reversible endothermic reactions.* Chemical Engineering Science, 57, 855-872.

Berezowski, M. (2010). *The application of the parametric continuation method for determining steady state diagrams in chemical engineering.* Chemical Engineering Science, 65/19, 5411-5414.

Berezowski, M. (2011). *Bifurcation analysis of the conversion degree in systems based on the cascade of tank reactors.* Chemical Engineering Science, 66/21, 5219-5223, 2011.

Berezowski, M. & Kulik, B. (2009). *Periodicity of chaotic solutions of the model of thermally coupled cascades of chemical tank reactors with flow reversal.* Chaos, Solitons & Fractals, 40/1, 331-336.

Glöckler, B., Kolios, G. & Eigenberger, G. (2003). *Analysis of a novel reverse-flow reactor concept for autothermal methane steam reforming.* Chemical Engineering Science, 58, 593-601.

Gupta, V.K. & Bhata, S.K. (1991). *Solution of cyclic profiles in catalytic reactor operation with periodic flow reversal.* Comp. Chem. Engng., 15/4, 229-237.

Jeong, Y.O. & Luss, D. (2003). *Pollutant destruction in a reverse-flow chromatographic reactor.* Chemical Engineering Science, 58, 1095-1102.

Knihast, J., Jeong, Y.O. & Luss, D. (1999). Dependence of cooled reverse-flow reactor dynamics on reactor model. *AIChE J.,* 45/2, 299-309.

Kulik, B. & Berezowski, M. (2008). *Chaotic dynamics of coupled cascades of tank reactors with flow reversal,* Chemical and Process Engineering, 29, 465-471.

Purwono, S., Budman, H., Hudgins, R.R., Silveston, P.L. & Matros, Yu.Sh. (1994). *Runway in packed bed reactors operating with periodic flow reversal.* Chemical Engineering Science, 49/24B, 5473-5487.

53

Řeháček, J., Kubiček, M. & Marek. M. (1998). *Periodic, quasiperiodic and chaoic spatiotemporal patterns in a tubular catalytic reactor with periodic flow reversal.* Comp. Chem. Engng., 22/1-2, 283-297.

Sheintuch, M. (2005). *Analysis of design sensitivity of flow-reversal reactors: Simulations, approximations and oxidations experiments.* Chemical Engineering Science, 60, 2991-2998.

Sheintuch, M. & Nekhamkina, O. (2004). *Comparison of flow-reversal, internal-recirculation and loop reactors.* Chemical Engineering Science, 59, 4065-4072.

Żukowski, W. & Berezowski, M. (2000). *Generation of chaotic oscillations in a system with flow reversal.* Chemical Engineering Science, 55, 339-343.

Chapter 6. Crisis phenomenom

Grebogi, Romeiras, and Yorke distinguished between three types of crises (Grebogi et. al, 1983). The first type, a boundary or an exterior crisis, the attractor is suddenly destroyed as the parameters are varied. In the second type of crisis, an interior crisis, the size of the chaotic attractor suddenly increases. In the third type, an attractor merging crisis, two or more chaotic attractors merge to form a single attractor as the critical parameter value is passed.

This chapter is focused on a scenario of generating the second type of crisis on the grounds of the amplitude spectrum of a reactor model (Berezowski, 2013).

Tubular chemical reactors with mass recycle are commonly used in industrial processes. Due to feedback which is effected by recycle, certain complex static and dynamical phenomena may arise (Luss & Amundson, 1967; Reilly & Schmitz, 1966; Reilly & Schmitz, 1967; Berezowski, 1990; Berezowski, 1993) including chaos (Jacobsen & Berezowski, 1998; Berezowski, 2000; Berezowski, 2001; Antoniades & Christofides, 2000; Antoniades & Christofides, 2001; Berezowski, 2006).

The dynamics of homogeneous non-adiabatic tubular chemical reactor with mass recycle is subjected to analysis, with particular focus on the crisis phenomenon which occurs during chaotic oscillations of the reactor. Assuming single reaction of the $A \to B$ type of arbitrary order, the balance equations may be expressed in the following dimensionless form (Jacobsen & Berezowski, 1998; Berezowski, 2006):

the mass balance:

$$\frac{\partial \alpha}{\partial \tau} + \frac{\partial \alpha}{\partial \xi} = (1 - f)\phi(\alpha, \Theta) \qquad (6.1)$$

the heat balance:

$$\frac{\partial \Theta}{\partial \tau} + \frac{\partial \Theta}{\partial \xi} = (1 - f)\phi(\alpha, \Theta) + (1 - f)\delta(\Theta_H - \Theta). \qquad (6.2)$$

Function ϕ describes the kinetics of a chemical reaction, expressed as:

$$\phi = Da(1 - \alpha)^m \exp\left(\gamma \frac{\beta \Theta}{1 + \beta \Theta}\right). \qquad (6.3)$$

The boundary conditions resulting from recycle assume the form:

$$\alpha(0, \tau) = f\alpha(1, \tau); \quad \Theta(0, \tau) = f\Theta(1, \tau). \qquad (6.4)$$

The above model offers diverse types of static and dynamical solutions, including: multiple steady states, limit cycles, quasiperiodic orbits or chaotic orbits – see: (Jacobsen & Berezowski, 1998; Berezowski, 2000; Berezowski 2001). A specific type of solution was shown in (Berezowski, 2003) on the bifurcation diagram in Fig.1.

For $\Theta_H = -0.0275$ a rapid growth of the variables value is noticeable. This phenomenon is labelled as *"crisis"*, as chaotic solutions can drastically change the amplitude, even at minimal change of the model parameter values. Thus, it is a bifurcation case which cannot be explained only by means of steady-state diagrams applicable for Fig.6.1 or the time series.

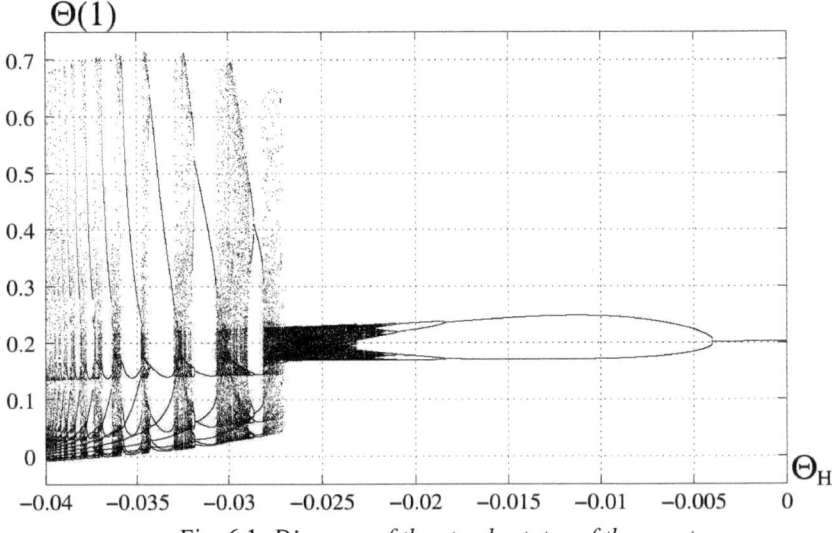

Fig. 6.1. *Diagram of the steady states of the reactor*

As already stated, the mechanism of the "*crisis*" phenomenon may be explained by means of the amplitude spectrum diagram (Fig. 6.2).

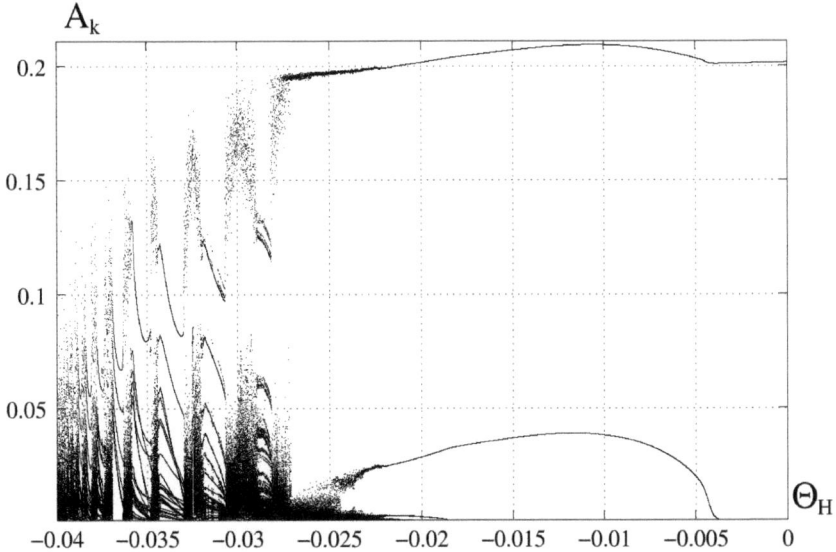

Fig. 6.2. *Diagram of the amplitude spectrum of the reactor*

The specific points correspond to the values of harmonic amplitudes. In consideration of discrete time graphs of temperature and concentration (Jacobsen & Berezowski, 1998), the amplitude spectra were derived from Fourier's discrete transform:

$$A_k = \frac{1}{N}\left|\sum_{n=0}^{N-1}\Theta_n(1)\exp\left(-i\frac{2\pi kn}{N}\right)\right| \qquad (6.5)$$

where: i – imaginary number, n – sample number, N – sample quantity, k – harmonic number, $\Theta_n(1)$ - values of dimensionless temperature at the reactor outlet at discrete moment of time "n".

From the point of view of the analysis concerning the crisis phenomenon the upper part of the amplitude spectrum in Fig.6.2 is important, as it shows the scenario of reaching the crisis involving mild decomposition of the upper branch of the diagram. The branch points correspond to the zero harmonic.

As far as the range of chaotic solutions is concerned, before the crisis (for example, when: $\Theta_H = -0.025$) the spectrum graph contains the zero harmonic with the dominant amplitude and infinite number of other harmonics resulting from chaos (Fig.6.3).

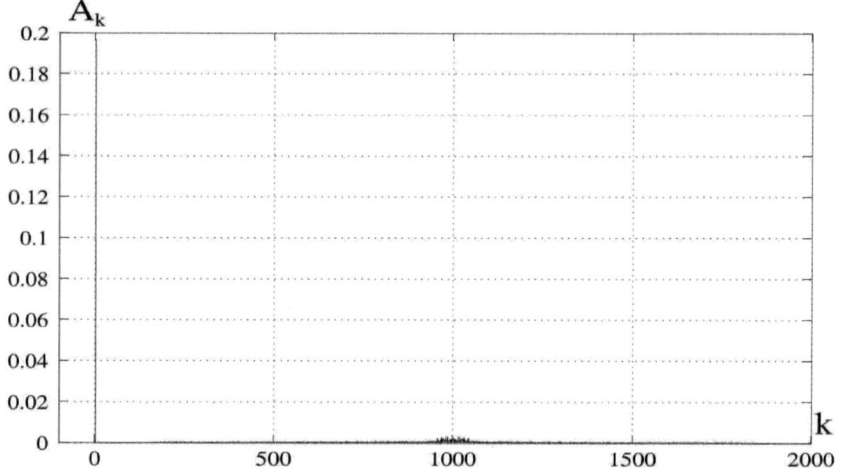

Fig. 6.3. *Amplitude spectrum of the reactor for* $\Theta_H = -0.025$

The time series of the dimensionless temperature referring to the discussed case is shown in Fig.6.4.

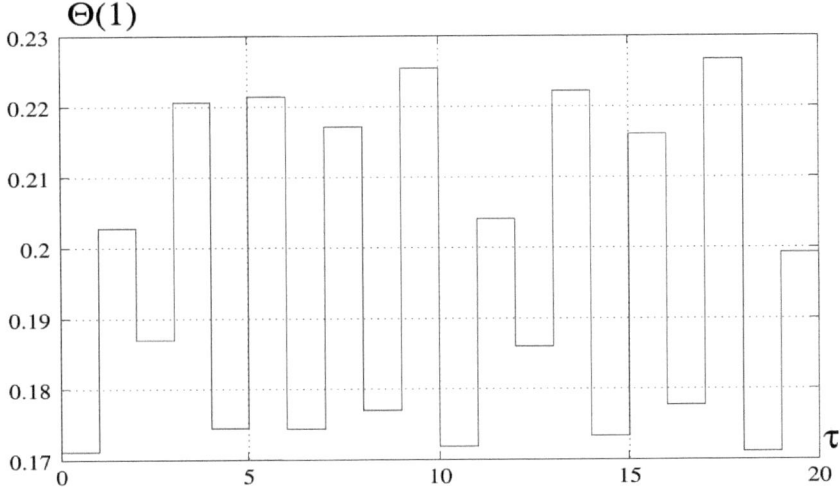

Fig. 6.4. *Time series of the dimensionless temperature for* $\Theta_H = -0.025$

After entering the domain of crisis (for example, when: $\Theta_H = -0.03$) the character of the spectrum graph shows no qualitative changes (Fig.6.5).

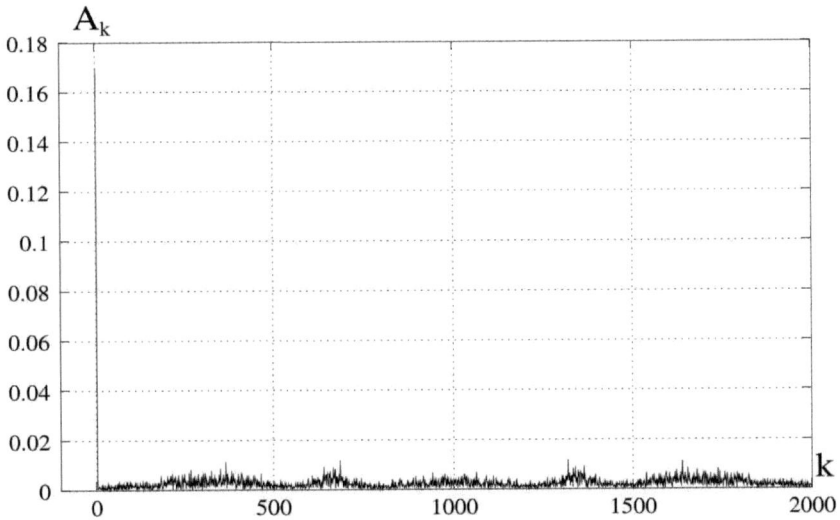

Fig. 6.5. *Amplitude spectrum of the reactor for* $\Theta_H = -0.03$

However, the quality of the time series changed, as high values amplitudes appeared (Fig.6.6).

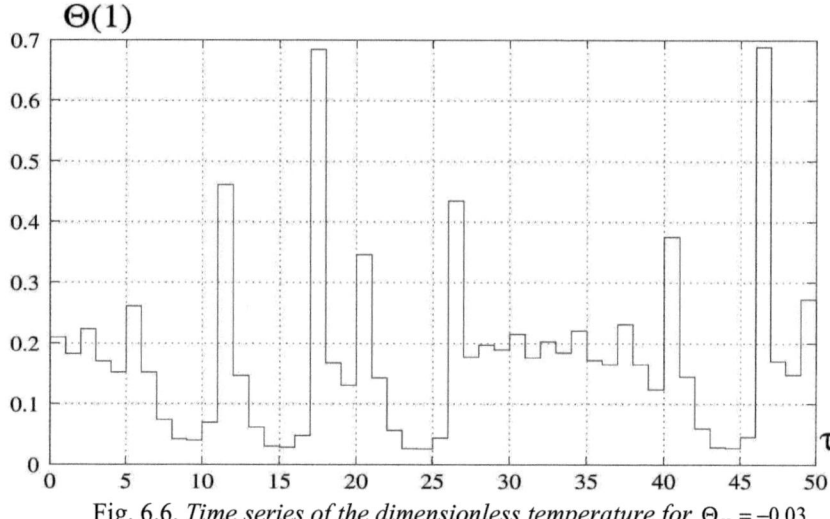

Fig. 6.6. *Time series of the dimensionless temperature for* $\Theta_H = -0.03$

Zero - harmonic is the average of the samples. As shown in Figure 6.2, 6.3 and 6.5, after passing through a crisis phase practically unchanged the value of this harmonic. This means that practically did not change the average temperature. Changed only the temperature distribution (Figures 6.4 and 6.6).

The following numerical values of the individual coefficients have been assumed in the computations: $n = 1.5$, $\gamma = 15$, $\beta = 2$, $\delta = 3$, $f = 0.5$, $Da = 0.15$.

Rerefences

Antoniades C. & Christofides P.D. (2000). *Nonlinear Feedback Control of Parabolic Partial Differential Difference Equation Systems*. International Journal of Control, 73, 1572-1591.

Antoniades C. & Christofides P.D. (2001). *Studies on Nonlinear Dynamics and Control of a Tubular Reactor with Recycle*. Nonlinear Analysis – Theory Methods and Applications, 47, 5933-5944.

Berezowski M. (1990). *A sufficient condition for the existence of single steady states in chemical reactors with recycle*. Chemical Engineering Science, 45, 1325-1329.

Berezowski M. (1993). *Dynamics profiles in chemical reactors with recycle*. Chemical Engineering Science, 48, 2799-2806.

Berezowski M. (2000). *Spatio-temporal chaos in tubular chemical reactors with the recycle of mass*. Chaos, Solitons & Fractals, *11*, 1197-1204.

Berezowski M. (2001). *Efect on delay time on the generation of chaos in continuous systems. One-dimensional model. Two-dimensional model – tubular chemical reactor with recycle*. Chaos, Solitons & Fractals, *12*, 83-89.

Berezowski M. (2003). *Fractal solutions of recirculation tubular chemical reactors*. Chaos, Solitons & Fractals, *16*, 1-12.

Berezowski M. (2006). *Fractal character of basin boundaries in a tubular chemical reactor with mass recycle*. Chemical Engineering Science, 61/4, 1342-1345.

Berezowski M. (2013). *Crisis phenomenon in a chemical reactor with recycle*. Chemical Engineering Science, Volume 101, 20 September 2013, Pages 451–453.

Grebogi C., Ott E. & Yorke J.A. (1983). *Crises, sudden changes in chaotic attractors and transient chaos*. Physica D, 7, 181–200.

Jacobsen E.W. & Berezowski M. (1998). *Chaotic dynamics in homogeneous tubular reactors with recycle*. Chemical Engineering Science, *53*, 4023 – 4029.

Luss D. & Amundson N.R. (1967). *Stability of loop reactors*. A.I.Ch.E. J., 13, 279-290.

Reilly M.J. & Schmitz R.A. (1966). *Dynamics of a tubular reactor with recycle. Part I*. A.I.Ch.E. J., 12, 153-161.

Reilly M.J. & Schmitz R.A. (1967). *Dynamics of a tubular reactor with recycle. Part II*. A.I.Ch.E. J., 13, 519-527.

Chapter 7. Spatio-temporal chaos

One of the systems commonly employed in chemical engineering is that with the recycle of mass and/or heat. Such a system enables the energy and matter to be recovered in an industrial process. Recycle loops encountered in reactive processes are based on recycling the unreacted feed after withdrawal of the product, recycling the heat generated in the system (through external or internal heat exchangers), or the simultaneous recycling of both. The present study deals with the last type of recycle (Fig. 1).

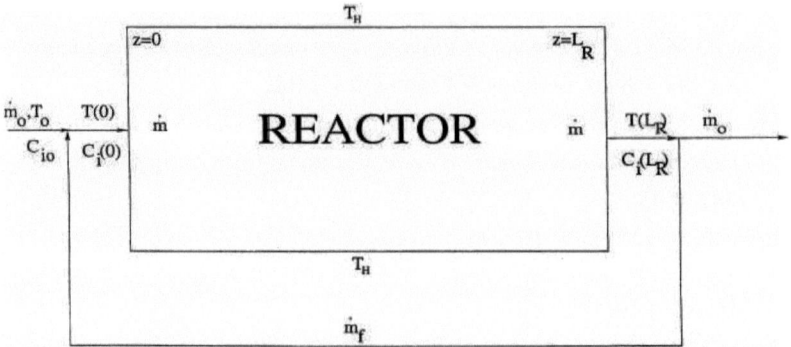

Fig. 7.1. Schematic diagram of a recycle reactor

For simplicity, we assume that a single reaction of the type $A \rightarrow B$ of any order occurs in the reactor and that the process is homogeneous. We also introduce the following definitions of the conversion degree:

$$\alpha = \frac{C_{A0} - C_A}{C_{A0}} \tag{7.1}$$

and the dimensionless temperature

$$\alpha = \frac{T - T_0}{\beta T_0}. \tag{7.2}$$

Hence, the mathematical model of the recycle reactor shown in Fig.7.1 is as follows:

$$\frac{\partial \alpha}{\partial \tau} + \frac{\partial \alpha}{\partial \xi} = (1 - f)\phi(\alpha, \Theta) \tag{7.3}$$

$$\frac{\partial \Theta}{\partial \tau} + \frac{\partial \Theta}{\partial \xi} = (1 - f)\phi(\alpha, \Theta) + (1 - f)\delta(\Theta_H - \Theta) \qquad (7.4)$$

The function / that describes the reaction kinetics is

$$\phi = Da(1 - \alpha)^m \exp\left(\gamma \frac{\beta \Theta}{1 + \beta \Theta}\right). \qquad (7.5)$$

The above model has to be supplemented with the following boundary conditions resulting from the presence of the recycle loop:

$$\alpha(0, \tau) = f\alpha(1, \tau) \qquad (7.6)$$

$$\Theta(0, \tau) = f\Theta(1, \tau) \qquad (7.7)$$

(it is assumed that the recycle delay is negligible, and also, that the mixing of the feed and recycle streams is instantaneous). There are a number of papers concerning the theoretical analysis of the statics and dynamics of a recycle reactor [1-3]; they all deal, however, with adiabatic systems. As will be shown in further discussion, qualitatively different results are obtained if an internal heat exchanger is introduced into the system, despite the fact that the temperature of the medium passing through the exchanger is assumed independent of the process conditions in the reactor itself. Reilly and Schmitz [9] and Luss and Amundson [8] were the first to attempt the analysis of such a system; their results are, however, far less comprehensive than those presented here (an overview of the conclusions has been given by Berezowski and Jacobsen [4] and Jacobsen and Berezowski [6]). The present analysis reveals a wide variety of dynamic phenomena, from uniperiodic to miltiperiodic to quasi-periodic oscillations, and also, chaotic spatio-temporal oscillations. These results are of tremendous practical importance, as they make it possible to avoid the regions characterized by intense oscillations already at the design stage. Chaotic oscillations can be detrimental to the process, though they may also have some positive characteristics, as pointed out recently by Chen-Chang-Chen [5] and Kopp et al. [7]: it is suggested that the oscillations should be induced on purpose in the system to determine the orbit of oscillations of a given frequency that would benefit the process. Assume that in the model (7.3)-(7.7) the recycle feedback is neglected ($f = 0$). The steady state in such a reactor is reached after the time equal to the residence time which, in dimensionless notation, means that $\tau_p = 1$. A simple numerical method for the solution of the system of partial differential equations can therefore be based upon a single integration of the di€erential Eqs. (3) and (4) from which the time derivatives were omitted. This procedure can thus be regarded as a particular form of the method of characteristics. The introduction of the recycle loop ($f > 0$) into the system after the time $\tau_p > 1$ leads to step changes in the inlet values $\alpha(0)$ and $\Theta(0)$, at the discrete moments $\tau_k = k$ (in the transient state or oscillatory state). These changes are carried towards the reactor outlet, and are detectable there at the moments $\tau_{k+1} = k+1$. Over the time interval $k < \tau < k+1$ both $\alpha(1)$ and $\Theta(1)$ remain constant. The profiles of the conversion degree $\alpha(\xi)$ and temperature $\Theta(\xi)$, at the moment τ_{k+1}, can thus be determined via a single integration of Eqs.(7.3) and (7.4) from which the time derivatives were removed. In its general form, the model (7.3)-(7.7) can be written as a discrete model [6]:

$$\frac{d\alpha_{k+1}}{d\xi} = (1 - f)\phi(\alpha_{k+1}, \Theta_{k+1}) \tag{7.8}$$

$$\frac{d\Theta}{d\xi} = (1 - f)\phi(\alpha_{k+1}, \Theta_{k+1}) + (1 - f)\delta(\Theta_H - \Theta_{k+1}) \tag{7.9}$$

$$\alpha_{k+1}(0, \tau) = f\alpha_k(1, \tau) \tag{7.10}$$

$$\Theta_{k+1}(0, \tau) = f\Theta_k(1, \tau) \tag{7.11}$$

A general solution to the above model, formulated for the reactor outlet, is

$$\begin{bmatrix} \alpha_{k+1}(1) \\ \Theta_{k+1}(1) \end{bmatrix} = F \begin{bmatrix} f\alpha_k(1) \\ f\Theta_k(1) \end{bmatrix}. \tag{7.12}$$

The system (7.12) is a discrete form. Consequently, to analyse the reactor dynamics (stability) we can use criteria taken from the literature [10]. The only dificulty is due to the fact that the function $\underline{F}$ is a certain integral functional rather than an algebraic relation, which is of crucial importance in determining the eigenvalues of the linearized model. However, since a single passage of the stream through the reactor constitutes a basic cycle period in the system $\tau_p = 1$, the eigenvalues of the linear approximation (7.8)-(7.11) can be evaluated as the eigenvalues of the monodromy matrix (also termed the ``circuit matrix", cf. [10]). This matrix is a solution to the following matrix di€erential equation:

$$D_\xi \underline{M} = \underline{A} \underline{M}. \tag{7.13}$$

The Jacobi matrix $\underline{A}$ is actualized at each integration step of Eq.(7.13). This means that Eqs. (7.8), (7.9) and (7.13) have to be integrated simultaneously. The initial condition associated with Eq.(7.13) is $\underline{M}(0) = f\underline{I}$ (in analysing the dynamics of multiple cycles the matrix $\underline{M}$ is determined through repeated integration of Eq.(7.13)). As already mentioned, the analysis carried out in this section has so far been based on the assumption that at any discrete moment $\tau_k = k$ the concentration and temperature profiles in the reactor are stationary and result from the solution of the ordinary di€erential Eqs.(7.8) and (7.9). Assume, though, that at a moment τ the value of any model parameter changed. Obviously, following this change the existing profiles will no longer be solutions to the set (7.8) and (7.9) and, over the time interval $k < \tau < k+1$, the values of the state variables at the reactor outlet $\alpha(1)$ and $\Theta(1)$, will not be constant. These values will undergo a continuous change, though at discrete moments $\tau_k = k$ the stepwise nature of the change will be preserved. Therefore, the analysis of the dynamics of the process thus defined should be based on the complete differential model (7.3) and (7.4), the time derivatives included. The model is solved employing the method of characteristics. If the coolant temperature, Θ_H, and the recycle coeffcient, f, are assumed as two principal bifurcation parameters of the model, the regions bounded by the appropriate bifurcation lines can be determined (Fig. 7.2).

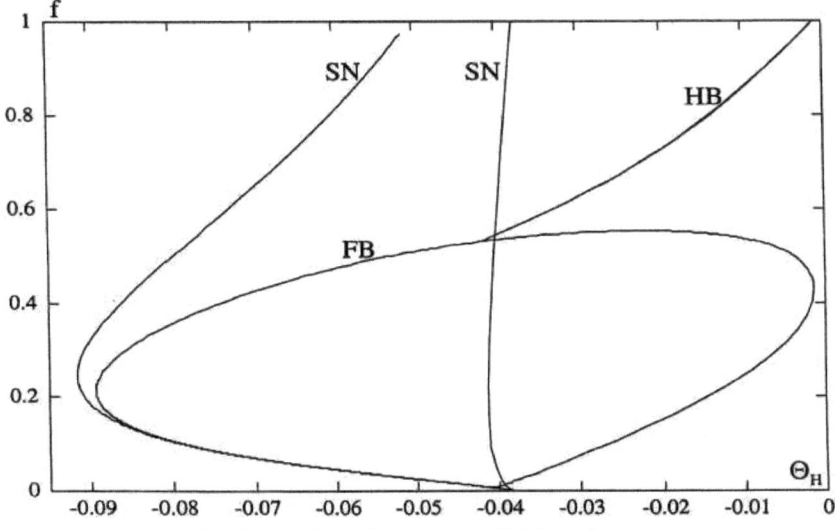

Fig. 7.2. Regions where three types of bifurcation can occur.

These lines are associated with the three types of bifurcation (static bifurcation - SN, flip bifurcation - FB [6] and Hopf bifurcation - HB). The values of the other parameters of the model are assumed as $Da = 0.15$, $\beta = 2$, $\delta = 3$, $n = 1.5$ and $\gamma = 15$. From Fig.7.2 it is seen that, for $f = 0.7$, there are two points of the static bifurcation SN ($\Theta_H = -0.066$ and $\Theta_H = -0.039$) and a single Hopf bifurcation point HB ($\Theta_H = -0.0222$) (the eigenvalue is complex and its modulus is one). This case is illustrated by the Feigenbaum diagram shown in Fig.7.3.

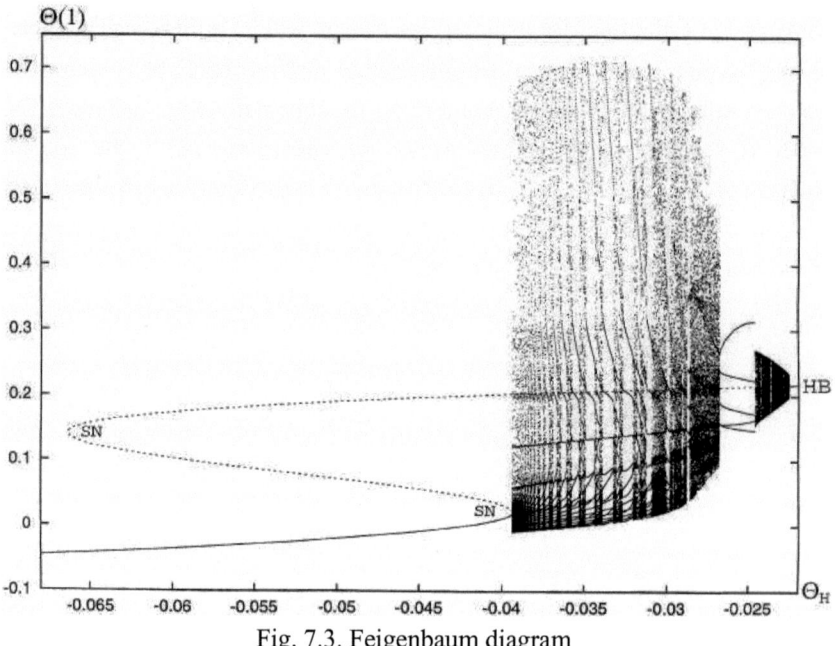

Fig. 7.3. Feigenbaum diagram

Quasi-periodic temporal solutions are clearly seen in Fig.7.3 (shaded area starting at HB). This shaded area is enlarged in Fig.7.4.

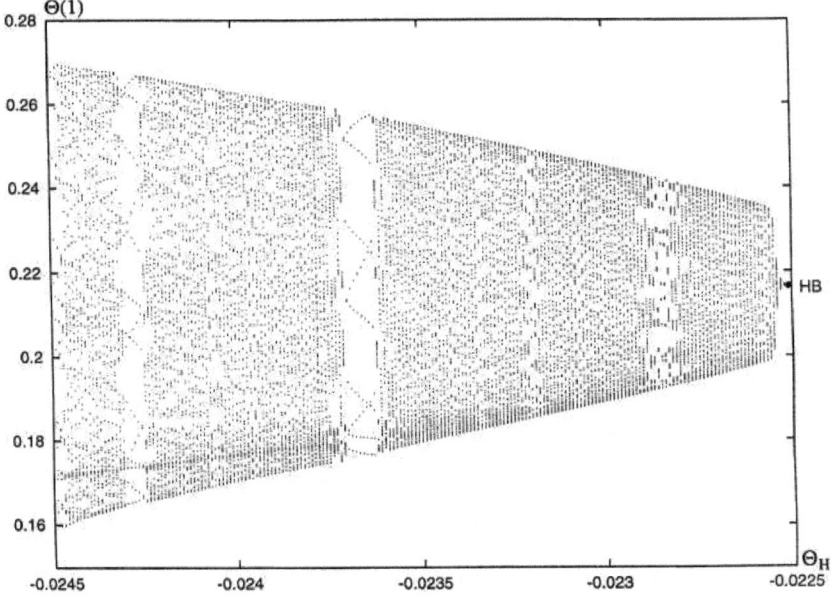

Fig. 7.4. Enlargement of a detail of Fig.7.3: region of quasi-periodic oscillations.

For comparison, Fig. 7.5(a) shows an enlargement of the chaotic region of Fig.7.3. The qualitative di€erences between the two cases are readily visible. The internal scenario of period doubling and intermittency, apparent in Fig.7.5a, is obviously missing from the quasi-periodic diagram (Fig.7.4). In the Feigenbaum diagram, Liapunov's exponent λ is plotted as a function of the bifurcation parameter Θ_H (Fig.7.5b).

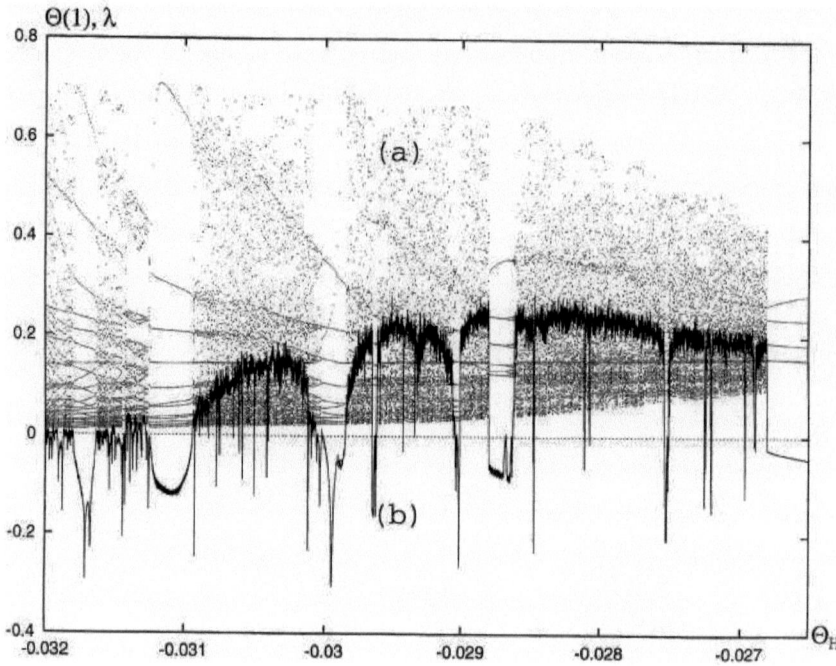

Fig.7.5. (a) Enlargement of a detail of Fig.7.3: region of chaotic oscillations. (b) Liapunov's exponent vs. bifurcation parameter Θ_H

It is readily seen that this exponent becomes zero at the bifurcation points. A positive value of λ indicates chaos. The procedure for determining λ is given in Appendix A7. One of the characteristics which makes it possible to differentiate between the quasi-periodic and chaotic oscillations is the insensitivity of the quasi-periodic signal to initial conditions (Fig.7.6).

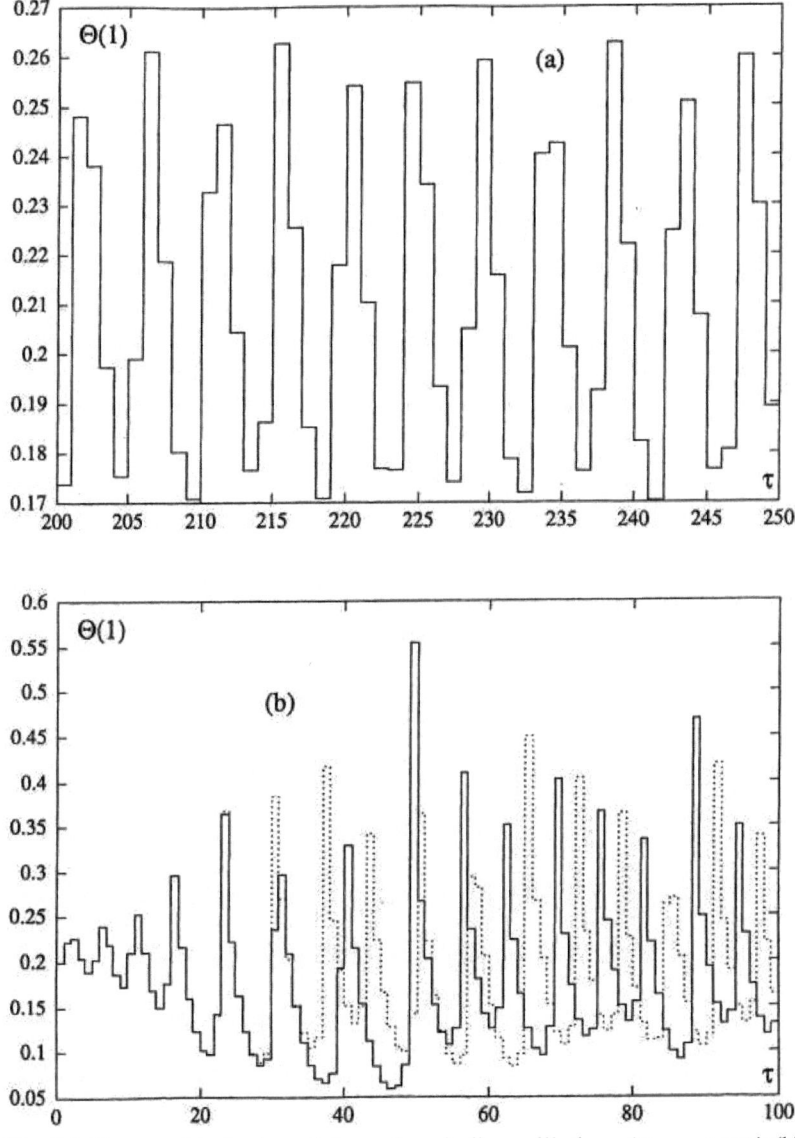

Fig.7.6. Temporal trajectory: (a) quasi-periodic oscillations ($\Theta_H = -0.024$);(b) chaotic oscillations ($\Theta_H = -0.028$ sensitivity on the initial conditions)

Also, the Poincare map looks altogether different. For the case of quasi-periodic oscillations the Poincare map is depicted in Fig.7.7.

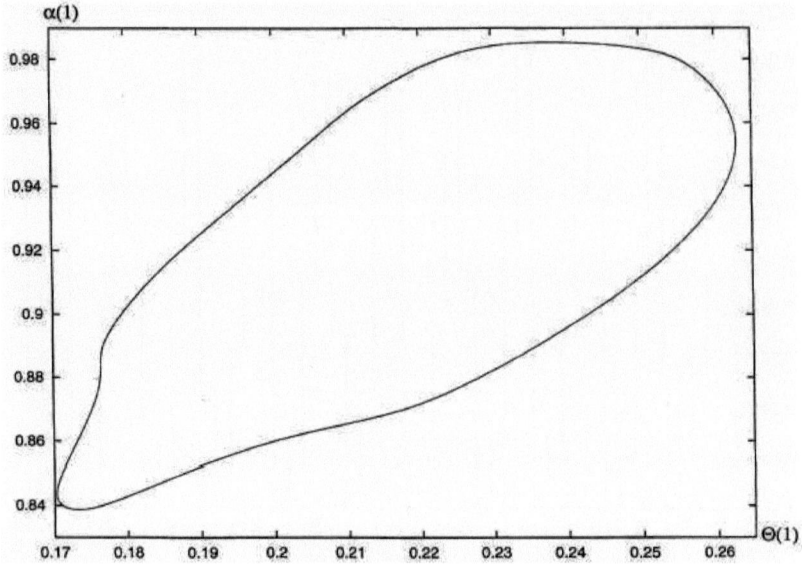

Fig.7.7. Poincare section for the case of quasi-periodic oscillations ($\Theta_H = -0.024$)

For the case of chaos, the Poincare map is shown in Fig.7.8.

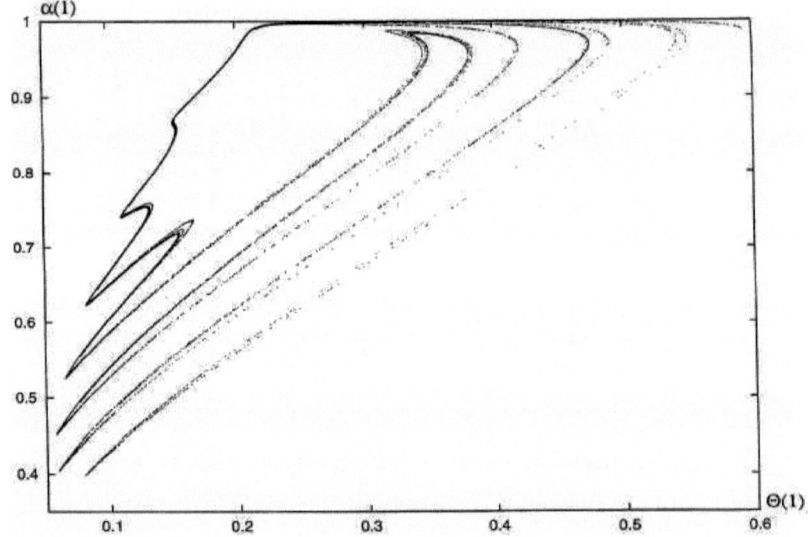

Fig.7.8. Poincare section for the case of chaotic oscillations ($\Theta_H = -0.028$)

The graph in this figure clearly has the shape of an Henon attractor, typical of discrete systems. In Figs.7.9 and 7.10 the so-called graphical iteration is shown of the dynamic solutions of the model. Fig.7.9 concerns the quasi-periodic state, whereas Fig.7.10 deals with chaos. The function F_2, presented in the graph, is a component of vector (7.12) and concerns the temperature.

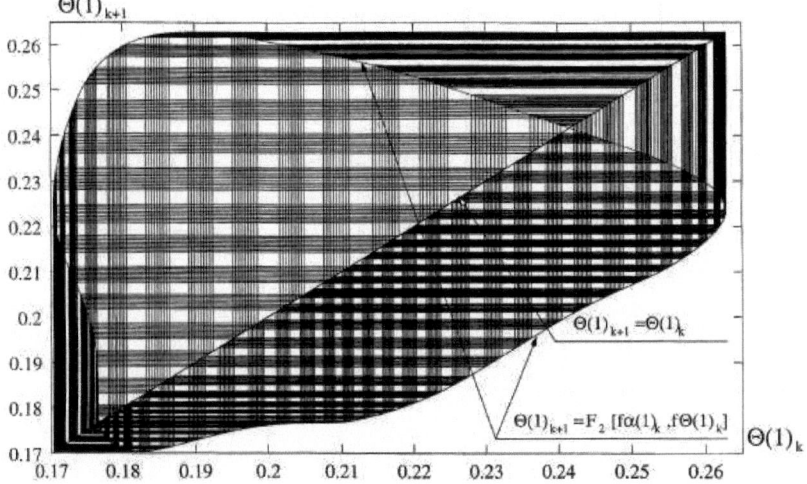

Fig.7.9. Graphical iteration of the quasi-periodic trajectory ($\Theta_H = -0.024$)

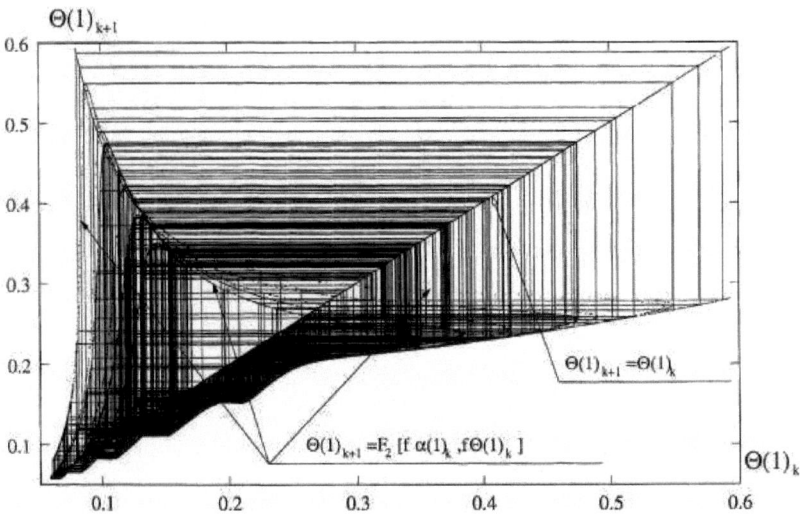

Fig.7.10. Graphical iteration of the chaotic trajectory ($\Theta_H = -0.028$)

The analysis carried out so far has been based on the assumption that the recycle loop becomes incorporated into the system after the time $\tau = 1$, that is, after at least one pass of the stream through the reactor. It has also been assumed that the values of all the model parameters are constant during the process. These restrictions will now be lifted, and the parameter values will be allowed to vary at any moment τ. As already mentioned, the instantaneous concentration and temperature profiles will then no longer have a stationary character and will fluctuate over the time interval $k < \tau < k+1$. It is further assumed that the cooling medium temperature, Θ_H, varies in the calculations. The simulation of the process dynamics requires the complete di€erential model, including the temporal partial derivatives. The balance equations are integrated using the method of characteristics. Representative results are shown in Fig. 11.

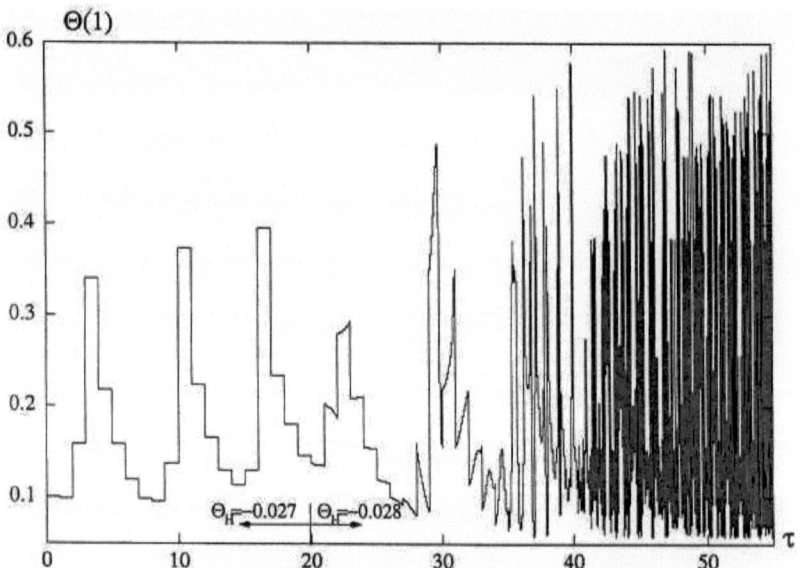

Fig.7.11. Example of the perturbation of Θ_H for the case of chaotic oscillations
$(\Theta_H = -0.027 \rightarrow -0.028)$

Until $\tau = 20$, Θ_H is -0.027, and a chaotic stepwise profile of the signal can be observed. At the time $\tau = 20$, Θ_H changes to -0.028. As can be seen from Fig.7.11, the temporal signal becomes totally deformed relative to that observed before the perturbation was introduced.

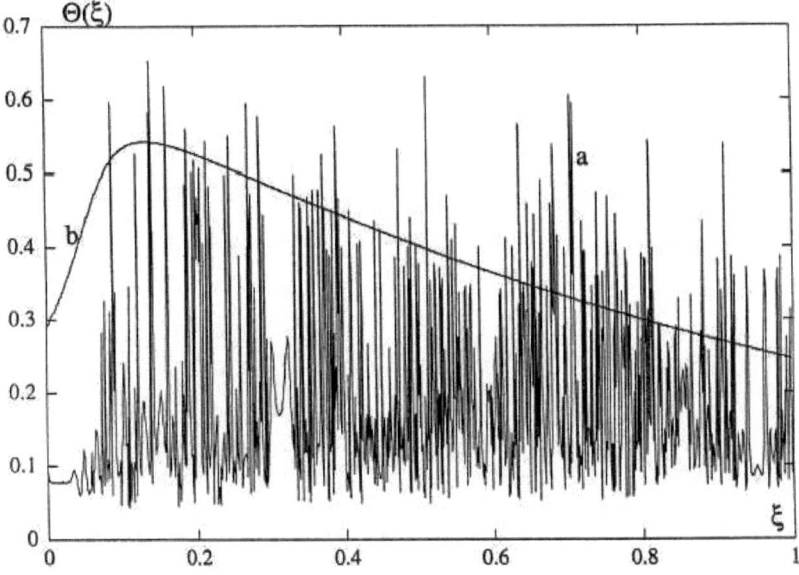

Fig.7.12. Temperature profiles at $\tau = 60$ ($\Theta_H = -0.027 \to -0.028$):(a) Θ_H perturbed; (b) Θ_H unperturbed

After a certain time has elapsed, the temporal changes become very rapid. Simultaneously, the profile of the state variable in the reactor exhibits rapid and steep spatial variations (Fig.7.12(a)). For the sake of comparison, the profile (b) is also drawn for $\Theta = -0.028$ but without any perturbations imposed on this parameter. In Fig.7.13 the spatial Poincare map is shown concerning a given moment τ. This is a typical cross-section of the chaotic attractor. It is worth noting that the envelope of this cross-section is formed by the Henon attractor of Fig.7.8.

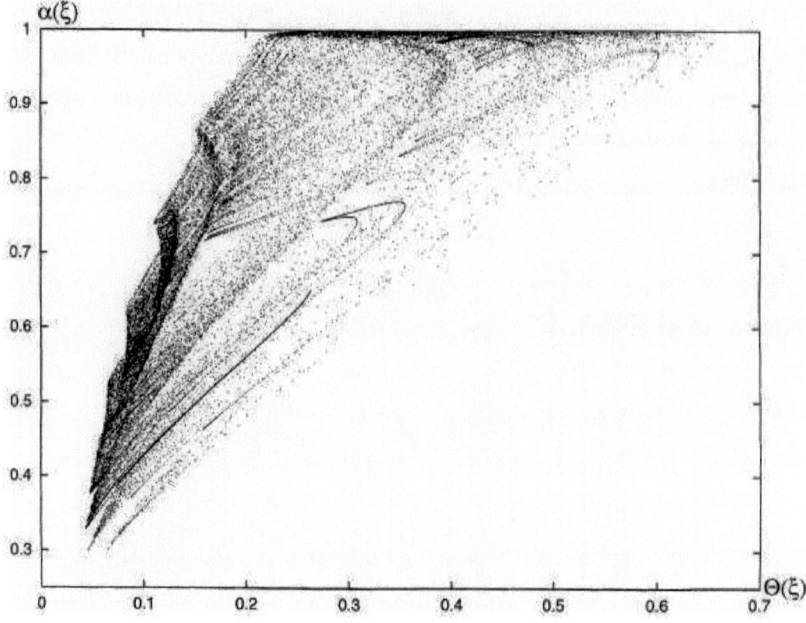

Fig.7.13. Spatial Poincare section at $\tau = 60$ ($\Theta_H = -0.027 \rightarrow -0.028$)

Appendix A7.

The solution of Eq. (12) obtained using the monodrome matrix $\underline{\underline{M}}$ has, for a single cycle, the following form:

$$\begin{bmatrix} \alpha_{k+1}(1) \\ \Theta_{k+1}(1) \end{bmatrix} = \underline{\underline{M}}_k \begin{bmatrix} \alpha_k(0) \\ \Theta_k(0) \end{bmatrix} \qquad (A7.1)$$

where the matrix

$$\underline{\underline{M}}_k = \begin{bmatrix} m_{11,k} & m_{12,k} \\ m_{21,k} & m_{22,k} \end{bmatrix} \qquad (A7.2)$$

is calculated based on relation (13).

If we deal with an oscillatory $(N+1)$-periodic cycle, the solution becomes

$$\begin{bmatrix} \alpha_{k+1}(1) \\ \Theta_{k+1}(1) \end{bmatrix} = \prod_{k=0}^{N} \underline{\underline{M}}_k \begin{bmatrix} \alpha_0(0) \\ \Theta_0(0) \end{bmatrix} = \underline{\underline{M}}_{(N)} \begin{bmatrix} \alpha_k(0) \\ \Theta_k(0) \end{bmatrix}. \qquad (A7.3)$$

Indicative of the stability of the solution α_{N+1}, Θ_{N+1} (i.e., of its sensitivity to the initial conditions α_0, Θ_0) are the eigenvalues s_i of $\underline{\underline{M}}_{(N)}$. Thus, Liapunov's exponent can be determined as

$$\lambda = \frac{1}{N}\ln\left(\max_i|s_i|\right). \tag{A7.4}$$

Elements of the successive matrices $\underline{\underline{M}}_k$ can be evaluated by integrating an equation of the type (13)

$$D_\xi \underline{\underline{M}}_k = \underline{\underline{A}} \underline{\underline{M}}_k \tag{A7.5}$$

from $\xi = 0$ to $\xi = 1$ subject to the initial condition

$$\underline{\underline{M}}_k(0) = f\,\underline{\underline{I}} \tag{A7.6}$$

or numerically from (A7.1) as

$$m_{11,k} = \frac{\partial \alpha_{k+1}(1)}{\partial \alpha_k(0)} \approx \frac{\Delta \alpha_{k+1}(1)}{\Delta \alpha_k(0)} \tag{A7.7}$$

$$m_{12,k} = \frac{\partial \alpha_{k+1}(1)}{\partial \Theta_k(0)} \approx \frac{\Delta \alpha_{k+1}(1)}{\Delta \Theta_k(0)} \tag{A7.8}$$

$$m_{21,k} = \frac{\partial \Theta_{k+1}(1)}{\partial \alpha_k(0)} \approx \frac{\Delta \Theta_{k+1}(1)}{\Delta \alpha_k(0)} \tag{A7.9}$$

$$m_{22,k} = \frac{\partial \Theta_{k+1}(1)}{\partial \Theta_k(0)} \approx \frac{\Delta \Theta_{k+1}(1)}{\Delta \Theta_k(0)} \tag{A7.10}$$

References

1. M. Berezowski, A. Burghardt, *A generalized analytical method for determining multiplicity features in chemical reactors with recycle*, Chem. Eng. Sci. 44 (1989) 2927.

2. M. Berezowski, *Method for analysing global stability of pseudohomogeneous chemical reactors with recycle*, Chem. Eng. Sci. 46 (1991) 1781.

3. M. Berezowski, *Dynamic profiles in chemical reactors with recycle*, Chem. Eng. Sci. 48 (1993) 2799.

4. M. Berezowski, E.W. Jacobsen, *On the dynamics of homogeneous tubular reactors with recycle*, AIChE Annual Meeting, Paper 266i, Los Angeles, 1997.

5. Chen Chang-Chen, *Stabilized chaotic dynamics of coupled nonisothermal CSTRs*, Chem. Eng. Sci. 51 (1996) 5159.

6. E.W. Jacobsen, M. Berezowski, *Chaotic dynamics in homogeneous tubular reactors with recycle*, Chem. Eng. Sci. 53 (1998) 4023.

7. T. Kopp, H. Zimmermann, Martienssen, *Steuerung und Analyse komplexer Systemzust ande, Statusseminar Nichtlineare Dynamik bei chemischen Prozessen*, Frankfurt, 41 1997.

8. D. Luss, N.R. Amundson, *Stability of loop reactors*, AIChE J. 13 (1967) 279.

9. M.J. Reilly, R.A. Schmitz, *Dynamics of a tubular reactor with recycle Part II*, AIChE J. 13 (1967) 519.

10. R. Seydel, *Practical bifurcation and stability analysis*, Springer, Berlin, 1994.

11. M. Berezowski, *Spatio-temporal chaos in tubular chemical reactors with the recycle of mass*. Chaos,Solitons&Fractals, **11**, 1197-1204, 2000.

Notations

A_q heat exchange area, m^2

Da Damköhler number $\left(= \dfrac{V_R(-r_0)}{\dot{F}C_{A0}} \right)$

E activation energy, $kJ/kmol$

f recycle fraction, $\left(= \dfrac{\dot{m}_f}{\dot{m}_R} \right)$

$\dot{F}$ volumetric flow rate, m^3/s

$(-\Delta H)$ heat of reaction, $kJ/kmol$

k reaction rate constant, $1/s\left(m^3/kmol\right)^{n-1}$

k_q heat exchange coefficient, $W/\left(m^2 K\right)$

L length, m

Le Lewis number, $\left(1 + \dfrac{m_w c_{pw}}{m_m c_{pm}} \right)$

$\dot{m}$ mass flow rate, kg/s

n order of reaction

$(-r)$ rate of reaction, $\left(= kC^n \right)$, $kmol/\left(m^3 s\right)$

R gas constant, $kJ/(kmol\ K)$

t	time, s
T	temperature, K
V	volume, m^3
z	position, m

Greek letters

α degree of conversion $\left(=\dfrac{C_{A0}-C_A}{C_{A0}}\right)$

β dimensionless number related to adiabatic temperature increase $\left(=\dfrac{(-\Delta H)C_{A0}}{T_0\rho c_p}\right)$

γ dimensionless number related to activation energy $\left(=\dfrac{E}{RT_0}\right)$

δ dimensionless heat exchange coefficient $\left(=\dfrac{A_q k_q}{\rho c_p \dot{F}}\right)$

Θ dimensionless temperature $\left(=\dfrac{T-T_0}{\beta T_0}\right)$

ξ dimensionless position $\left(=\dfrac{z}{L}\right)$

ρ density $\left(=\dfrac{kg}{m^3}\right)$

τ dimensionless time $\left(=\dfrac{\dot{F}}{V_R}t\right)$

τ_{rf} dimensionless reverse period

τ_{rel} dimensionless relaxation time